THE NATIONAL ACADEMIES
Advisers to the Nation on Science, Engineering, and Medicine

The **National Academy of Sciences** is a private, nonprofit, self-perpetuating society of distinguished scholars engaged in scientific and engineering research, dedicated to the furtherance of science and technology and to their use for the general welfare. Upon the authority of the charter granted to it by the Congress in 1863, the Academy has a mandate that requires it to advise the federal government on scientific and technical matters. Dr. Bruce M. Alberts is president of the National Academy of Sciences.

The **National Academy of Engineering** was established in 1964, under the charter of the National Academy of Sciences, as a parallel organization of outstanding engineers. It is autonomous in its administration and in the selection of its members, sharing with the National Academy of Sciences the responsibility for advising the federal government. The National Academy of Engineering also sponsors engineering programs aimed at meeting national needs, encourages education and research, and recognizes the superior achievements of engineers. Dr. Wm. A. Wulf is president of the National Academy of Engineering.

The **Institute of Medicine** was established in 1970 by the National Academy of Sciences to secure the services of eminent members of appropriate professions in the examination of policy matters pertaining to the health of the public. The Institute acts under the responsibility given to the National Academy of Sciences by its congressional charter to be an adviser to the federal government and, upon its own initiative, to identify issues of medical care, research, and education. Dr. Harvey V. Fineberg is president of the Institute of Medicine.

The **National Research Council** was organized by the National Academy of Sciences in 1916 to associate the broad community of science and technology with the Academy's purposes of furthering knowledge and advising the federal government. Functioning in accordance with general policies determined by the Academy, the Council has become the principal operating agency of both the National Academy of Sciences and the National Academy of Engineering in providing services to the government, the public, and the scientific and engineering communities. The Council is administered jointly by both Academies and the Institute of Medicine. Dr. Bruce M. Alberts and Dr. Wm. A. Wulf are chair and vice chair, respectively, of the National Research Council.

www.national-academies.org

Roundtable Staff

Christine M. Coussens, Study Director
Ricardo Molins, Senior Program Officer
Dalia Gilbert, Research Associate
Erin McCarville, Senior Project Assistant
Victoria Blaho, Christine Mirzayan Science and Technology Policy Graduate
    Fellowship Program Intern
LaTeya Foxx, Anderson Intern

Division Staff

Andrew Pope, Division Director
Troy Prince, Administrative Assistant
Carlos Gabriel, Financial Associate

DISASTERS ROUNDTABLE

Appointed Members

William H. Hooke (Chair), Senior Policy Fellow and Director of the Atmospheric Policy Program, American Meteorological Society, Washington, DC

Ross B. Corotis, Professor, Department of Civil Engineering, University of Colorado at Boulder, Boulder, CO

Ann-Margaret Esnard, Associate Professor, Director of GEDDeS GIS Lab, Department of City and Regional Planning, Cornell University, Ithaca, NY

Ellis M. Stanley, Sr., General Manager, City of Los Angeles Emergency Preparedness Department, Los Angeles, CA

Richard T. Sylves, Professor, Department of Political Science and International Relations, University of Delaware, Newark, DE

Susan Tubbesing, Executive Director, Earthquake Engineering Research Institute, Oakland, CA

Ex Officio Members

Stephen Ambrose, Physical Scientist, Earth Science Enterprise, National Aeronautics and Space Administration, Washington, DC

David Applegate, Ph.D., Senior Science Advisor for Earthquake and Geologic Hazards, U.S. Geological Survey, Reston, VA

Frank Best, Vice President, Alltech, Inc., Fairfax, VA

Lloyd S. Cluff, Manager, Geosciences, Pacific Gas and Electric, San Francisco, CA

Elizabeth Lemersal, Physical Scientist, Risk Assessment Branch, Mitigation Division, Federal Emergency Management Agency, Department of Homeland Security, Washington, DC

Dennis Wenger, Ph.D., Program Director, Infrastructure Management and Hazard Response, National Science Foundation, Arlington, VA

Helen Wood, Senior Advisor, Satellite and Information Services, National Oceanic and Atmospheric Administration, Washington, DC

Roundtable Staff

William A. Anderson, Associate Executive Director, Division on Earth and Life Studies, and Director, Disasters Roundtable

Patricia Jones Kershaw, Senior Program Associate, Disasters Roundtable

Byron Mason, Senior Project Assistant, Disasters Roundtable

Melissa Cole, Christine Mirzayan Science and Technology Policy Graduate Fellowship Program Intern, Disasters Roundtable

REVIEWERS

This report has been reviewed in draft form by individuals chosen for their diverse perspectives and technical expertise, in accordance with procedures approved by the National Research Council's Report Review Committee. The purpose of this independent review is to provide candid and critical comments that will assist the institution in making its published report as sound as possible and to ensure that the report meets institutional standards for objectivity, evidence, and responsiveness to the study charge. The review comments and draft manuscript remain confidential to protect the integrity of the deliberative process. We wish to thank the following individuals for their review of this report:

John Godleski, Harvard University, Boston, MA
John Harrald, George Washington University, Washington, DC
Havidán Rodríguez, University of Delaware, Newark, DE
Nate Szejniuk, University of North Carolina, Chapel Hill, NC

Although the reviewers listed above have provided many constructive comments and suggestions, they were not asked to endorse the final draft of the report before its release. The review of this report was overseen by **Melvin Worth**, Scholar-in-Residence, Institute of Medicine, who was responsible for making certain that an independent examination of this report was carried out in accordance with institutional procedures and that all review comments were carefully considered. Responsibility for the final content of this report rests entirely with the authoring committee and the institution.

Preface

The National Research Council's Disasters Roundtable and the Institute of Medicine's Roundtable on Environmental Health Sciences, Research, and Medicine were established as mechanisms for bringing various stakeholders together to discuss timely issues in a neutral setting. The goal was not to resolve these issues, but to create an environment conducive to scientific debate. The members of the respective Roundtables comprise representatives from academia, industry, nongovernmental agencies, and government, whose perspectives range widely and represent the diverse viewpoints of researchers, federal officials, and public interest.

This workshop was convened by the two Roundtables as a contribution to the debate on the health risks of disasters and the related need to build capacity to deal with them. The meeting was strengthened by integrating perspectives from these two fields, so that the agenda represented information from both communities and provided an opportunity to look at some of the most pressing research and preparedness needs for health risks of disasters.

Disasters, almost by definition, involve health risks, and have frequently been associated with first responders saving lives in the face of extreme events such as hurricanes, earthquakes, or flooding by transporting injured victims to hospitals to receive care. Life then continues until the next disaster arrives. Perceptions changed, however, with the terrorist attacks of September 11 and the subsequent anthrax attacks, when the government and the public realized the need for more attention to the complex health risks associated with disasters. More emphasis has also been placed on long-term needs after disasters as recovery continues long after release from the hospital or the burying of the dead. In short, what is clear is that preparing for health risks must occur long before disasters strike, and addressing health problems continues long after the initial "search and rescue" and other emergency period activities.

Since 2001, there has been a greater need for integrated, up-to-date scientific information to respond to the rapidly changing circumstances that occur with

disasters. Significant strides toward integration have occurred, but it is clear that additional planning, research, and integration are needed. Unlike many scientific subjects, where the practitioner's knowledge is solid, but public awareness lags, this is one area where professional understanding, capabilities, and approaches are evolving rapidly and substantially.

Current discussions of disasters tend to center on terrorist attacks and health risks. It is important to remember, however, that disasters are a multi-faceted challenge and include the public health consequences of geophysical hazards, industrial/technological accidents, terrorist events, and biological disasters, such as SARS outbreaks and *E. coli* contamination. In addition, on the international scale, disasters include the complex disasters resulting from war, government collapse, and famine. While September 11 caught the United States offguard, it is important that not all of our resources go into one area. We need to continue to have the ability to respond to a variety of threats.

Risk communication has become increasingly important as individuals receive information from various media (e.g., newspaper, television, radios, internet), and may seek to validate their knowledge with local experts, trusted friends, and personal experience. With the advent of 24-hour news coverage and the desire for up-to-date information, there are new challenges for risk communication. While it is important that messages from the government be consistent across agencies, it is also important that the messages be clear and honest, while not understating the risks. Scientists and policy makers need to build on the strength of the established literature of risk communication to fill in the gaps that are important for disasters.

Personnel needs were discussed by many speakers throughout the day. The issues ranged from providing responders with ongoing training and information on health risks to replacing an aging workforce. Training will continue to be important as the disasters that we are likely to face on a national scale will involve many complex problems. Training will need to be both general and specific, because the type, the magnitude, and the timing of the threats are unknown. How to prepare for integration in a climate of uncertainty is an area of ongoing discussion.

This workshop summary captures the discussions and presentations by the speakers and participants, who identified the areas in which additional research is needed, the processes by which changes can occur, and the gaps in our knowledge. The views expressed here do not necessarily reflect those of the National Research Council, the Institute of Medicine, the Roundtables, or their sponsors.

Paul G. Rogers
Chair
Roundtable on Environmental Health
Sciences, Research, and Medicine

William H. Hooke
Chair
Disasters Roundtable

Contents

Summary[1]

The National Research Council's Disasters Roundtable and the Institute of Medicine's Roundtable on Environmental Health Sciences, Research, and Medicine were formed to provide a neutral setting for individuals with different backgrounds and perspectives to discuss sensitive issues of mutual interest. Both groups bring together participants from the academic community, government, and the private sector who are actively engage in the disasters field (Disaster Roundtable) or environmental health sciences (Roundtable on Environmental Health Sciences, Research, and Medicine). Through their discussions, the Roundtables help to identify both current and potential problems, and consider approaches to solve them. The aim of these discussions is to share knowledge and ideas, but not proffer formal advice or recommendations.

This particular workshop provided an opportunity for the stakeholders in the two Roundtables to gather and consider issues related to health risks of disasters. To explore the capacity needs for addressing health risk during disasters, the speakers, participants, and Roundtable members considered how the United States will rise to meet these challenges and what research and training priorities were needed to strengthen its response to health-related risks.

INTERDISCIPLINARY PREPAREDNESS AND RESPONSE PLANS

Without a precise metric for preparedness, readiness can never be guaranteed. As a result, workshop participants stressed the need for enhanced collaboration and coordination among all stakeholders involved in disaster preparedness and response, in order to translate current policies into more concrete and effective

---

[1]This chapter was prepared by Melissa Cole from the transcript of the meeting. The discussions were edited and organized around major themes to provide a more readable summary and to eliminate duplication of topics.

response strategies. Traditional response efforts have relied upon local, state, and federal resources. In addition to strengthening coordination at all levels of government, workshop participants advocated expanding preparation, mitigation, and response efforts to include hospitals, health care professionals from all fields of social and traditional medicine, non-governmental organizations, mass media, private businesses, academia, and the engineering and scientific communities.

Recognizing the need for a unified approach to preparedness, the Department of Homeland Security (DHS) has recently developed the National Response Plan to improve coordination between government agencies and local first responders, noted Lew Stringer, of the Department of Homeland Security. Although not fully implemented as of the time of the workshop, this all-hazards plan addresses prevention, preparedness, response, and recovery for all levels of domestic incident management. Under the National Response Plan, local jurisdictions will retain primary responsibility for response efforts, using locally available resources; however, in the event of a large-scale catastrophe, local and state resources are likely to be overwhelmed. The Catastrophic Incident Plan, a supplement to the National Response Plan, has also been drafted to be immediately implemented during crisis by the DHS Secretary. The major goal of the plan is to provide accelerated deployment of federal assets to disaster zones. Pharmaceuticals and medical supplies from the Strategic National Stockpile and personnel from the U.S. Public Health Service Commissioned Corps Readiness Force, the Department of Veterans Affairs, the Department of Defense, and the National Disaster Medical System can reach disaster zones within twelve hours following a decision to deploy. While those assets will certainly help to augment the local response, noted Stringer, there is a need for another 20,000 trained and credentialed response personnel, in addition to the existing VA, USPHS, Department of Defense, and National Disaster Medical System (NDMS) staffs, to stage an effective mass-casualty response.

Enhancing local response capabilities through federal assets is only one example of creating a multi-level response. The Department of Health and Human Services has taken this a step further by incorporating both hospital and public health preparedness standards into their emergency preparedness grants, thereby emphasizing the importance of integrating health care systems' response plans with local jurisdictions' plans. According to William Raub, of the Department of Health and Human Services, the goals of the preparedness grants are to improve the nation's response capabilities in bioterrorism and other disasters, while correcting decades of neglect in the public health infrastructure.

In order to bring the full range of the nation's preparedness capabilities to bear, Jack Azar, of the Xerox Corporation, advocated the value of including the private sector in emergency response planning. In the event that a disaster occurs during regular business hours, business and industry executives must have updated and well-exercised plans, including evacuation and shelter-in-place

protocols, for protecting America's 100 million workers. According to Azar, the opportunity for government officials and private executives to share the successes and failures of emergency response, crisis management, and business continuity plans would be beneficial for all parties involved. That is especially true, considering the likelihood that private sector employees will be needed to assist in local response efforts. Following the September 11 attacks, engineers, iron workers, steamfitters, teamsters, electrical workers, and other building and construction trades unions collaborated with first responders throughout rescue, recovery, and cleanup efforts. While the National Institute of Environmental Health Sciences' (NIEHS) Worker Education and Training Program made a considerable effort to educate and train over 4,000 workers at Ground Zero, many workers, unfortunately, suffer from residual respiratory problems brought on by hazardous fumes at the site. Samuel Wilson, of the NIEHS, stressed the need for the academic and scientific communities to develop a standardized occupational safety framework for emergency responders that addresses issues like training, medical surveillance, protective equipment, and decontamination.

While many workshop participants discussed the importance of creating interdisciplinary, multi-level plans to respond to disasters, Rae Zimmerman, of New York University, emphasized addressing health risks through changes in engineering and infrastructure. Because of technological advances and economic necessity, much of the nation's infrastructure has become centralized and networked. For example, about 6.8 percent of all community water supply systems serve 45 percent of the population (Zimmerman, 2004: 81, based on U.S. EPA, 2002). In addition, much of this infrastructure is interconnected, for example, according to U.S. Geological Survey data (2004), electric power and thermal electric power plants consume roughly half the total water used in the United States. Thus, it is plausible that a breakdown in one component of the physical infrastructure could lead to cascading and escalating effects in other sectors. According to Zimmerman, future government and industry decisions must involve decoupling structural infrastructure and introducing flexibility into systems when repairs or alterations are necessary.

COMMUNICATING PREVENTION AND PREPAREDNESS TO THE PUBLIC

Once developed, preparedness and response plans must be communicated to affected communities, the general public, the scientific community, and other stakeholders to provide the information necessary to make the best possible decisions concerning their survival. Empathetic, accurate, and rapid emergency risk communication to culturally diverse audiences with variable levels of scientific literacy is critical for any preparedness and response effort. According to Julie Gerberding of the Centers for Disease Control and Prevention, emergency com-

munication messages are judged by their timeliness, content, and credibility and must imply an understanding of the range of emotions that affected individuals may experience.

Federal agencies, the media, and non-governmental organizations all play integral roles in disseminating risk communication messages. Broadcast media are the fastest and most widespread method for circulating important public health information during crises; therefore, working effectively with the media is essential to successful communication and response. On the other hand, members of the media may lack the background knowledge to immediately understand the scientific or technical issues involved in many disasters; therefore, Gerberding noted the importance of educating journalists so as to avoid misinformation.

Although the media has an expedient emergency broadcast system, Rocky Lopes of the American Red Cross asserted that, during crises, individuals want consistent messages from a variety of sources. That is especially true in light of the varying degrees of public trust in the United States government. As a non-governmental organization, the American Red Cross has noted that forty-eight percent of the public have said that they turn to the American Red Cross for disaster-related information. Through its collaborative efforts with the Federal Emergency Management Agency, the Department of Homeland Security, and the National Weather Service, the American Red Cross' process of verification and repetition reinforces messages and inspires community action.

VULNERABLE POPULATIONS

Effective risk communication messages can also mitigate the effects of disasters among those populations most vulnerable to their effects. During the 1995 Chicago heat wave, approximately 700 people, primarily elderly and poor residents, died in just three days. According to Eric Klinenberg of New York University, American society has assigned these populations less social importance, and this contributed to their isolation. As a result, their access to warnings, life-saving social interactions, and medical treatments was limited.

Following the Chicago heat wave, the Mayor's office implemented automated telephone heat warnings, targeting the elderly population. In addition, the city government began to work with the National Weather Service, private meteorologists, and community organizations to improve early detection of extreme weather, and to determine a graded series of warnings to be issued on television, radio, and in newspapers. While those were notable improvements, Klinenberg emphasized the value of reversing the societal trends of isolation and deprivation, which not only intensify fatalities during heat waves, but can accelerate fatalities in other crises, as well.

Children are another special population that is especially vulnerable to the health effects of disasters. J. R. Thomas, of the Franklin County, Ohio, Emergency Management Office, described children's different medical, legal, physical, and

psychological needs following a catastrophe. Medically, children need special supplies, such as pediatric drug doses and surgical instruments. In addition, emergency managers must plan for adequate shelter, transportation, and legal services to secure appropriate temporary placement for children who have been displaced from their families during the disaster. Even without parental separation, the traumatic experience of disasters alone is sufficient to produce psychological symptoms in children. Survivors of catastrophic events often have difficulty coherently verbalizing the effects of the disaster upon them. In young children, this is often compounded by an undeveloped language capacity. In addition, without a centralized mental health care system for children in the United States, treatment and services are currently scattered throughout numerous systems: schools, state and local health departments, child welfare services, and primary health care providers. As a result, the needs of this voiceless population are often underserved. To better prepare for future disasters, Thomas advocated including child health professionals from all fields of social and traditional medicine into response planning.

Recognizing vulnerable populations in disasters will help to ensure a more complete response; however, considering the unique nature of each disaster, different populations may be more vulnerable to specific disasters. Therefore, according to Carol Rubin, of the National Center for Environmental Health, a rapid community-needs assessment must be conducted following each disaster to ensure that the vital and specific needs of all affected community members are being met. A rapid needs assessment is a low cost, statistically sound, population-based epidemiological tool that can be used following a disaster to provide emergency managers with accurate and reliable information about the needs of an affected community. The results facilitate evidence-based decisions and interventions, providing a more effective disaster response through targeted allocation of scarce resources.

Since September 11, 2001, the federal government has undertaken significant initiatives to strengthen America's state and local emergency preparedness and response systems. The improvements in the nation's risk communication strategies, and its enhanced capabilities to acquire, store, and distribute pharmaceuticals and medical supplies to the public cannot be disputed; however, workshop participants stressed the importance of addressing the gaps and shortfalls in current emergency management policies. A number of challenges continue to exist as pointed out by many of the speakers and the participants, including:

- The acknowledgment that disasters may destroy local health infrastructure when it is needed most (p. 23).
- The concern that the public health workforce is nearing retirement age; thus, there is a critical need for training the next generation of responders (p. 11).

- The capacity limitation of the NDMS, if deployed during a disaster, to be able to respond to treat 224 inpatients and 4500 outpatients per day (p. 39).
- The need to engage the private sector in preparedness planning and communication channels for access to information in order to safeguard individuals at work (p. 49-52).
- The need to plan for management of facilities and personnel during sustained crises (p. 47).

1

Linking Hazards and Public Health: Communication and Environmental Health[1]

Disasters are the destructive forces that overwhelm a given region or community. These disasters can be natural or human-induced and require external assistance and coordination of services in order to address the myriad of effects and needs, including housing needs, transportation disruption, and health care needs. Disasters pose a variety of health risks, including physical injury, premature death, increased risk of communicable diseases, and psychological effects such as anxiety, neuroses, and depression. Destruction of local health infrastructure—hospitals, doctor's offices, clinics—is also likely to impact the delivery of health care services. A second wave of health care needs may occur due to food and water shortages and shifts of large populations to other areas.

In order to better understand the capacity needs for addressing health needs during disasters, the National Research Council's Disasters Roundtable and the Institute of Medicine's Roundtable on Environmental Health Sciences, Research, and Medicine sponsored a workshop on capacity needs during disasters. The summary has been prepared by the Roundtables' staff as the rapporteur to convey the essentials of the day's events. It should not be construed as a statement of the Roundtables—which can illuminate issues but cannot actually resolve them—or as a consensus study of The National Academies.

PUBLIC HEALTH RISKS ASSOCIATED WITH DISASTERS

After terrorists attacked the World Trade Center and the Pentagon on September 11, 2001, and anthrax was spread via the United States Postal Service only a month later, Americans felt ill prepared to respond to crises and to protect

---

[1]This chapter was prepared by Melissa Cole from the transcript of the meeting. The discussions were edited and organized around major themes to provide a more readable summary and to eliminate duplication of topics.

their health and well-being in the event of future attacks. Since then, preparedness activities have generated substantial interest and funding, and, as a result, federal, state, and local leaders are changing practices to prepare to respond to both natural and terrorist disasters; however, the improvements made are not nearly sufficient, noted some participants.

Usually politically motivated, the immediate goal of terrorism is to instill fear and confusion among the public. Immediately following an attack, the public's fear is transformed into intense preparation for the next crisis; yet, with increasing periods of safety, the public's sense of complacency tends to trump the preparedness activities, and, according to Dr. Julie Gerberding, Director of the Centers for Disease Control and Prevention, complacency is the enemy of public health. The public health and emergency management communities, therefore, have been charged with the task of conducting environmental analyses, educating and motivating the public to prepare themselves to mitigate the impacts of the next disaster, and communicating preparedness measures to the public before, during, and after an event. During the discussion, Gerberding reiterated that the public will need to become accustomed to the ideas that preparedness is not all or none. There can always be the potential for a scenario that is one step beyond the current level of preparedness. She further noted that this requires an ongoing sustained investment over time.

THE ROLE OF ENVIRONMENTAL HEALTH
IN UNDERSTANDING TERRORISM

Prior to designing disaster prevention and response strategies, it is critical to understand the physical and social environment surrounding terror agents. According to Dr. Lynn Goldman, The Johns Hopkins University Bloomberg School of Public Health, responders to a biological, chemical, or physical attack must be able to determine the following: where the agent is in the environment, where it will spread, who will be exposed, what quantity of the agent to which the victims may be exposed, what will happen to the exposed, what must be done to reduce exposure, and how to best treat victims. To answer those questions, it is necessary to understand the harmful agent, how it reaches the human body, and the health effects that it has on the body.

As Goldman noted, while conventional bombs have accounted for 46 percent of international terror attacks between 1963 and 1993, and 76 percent of domestic terror attacks between 1982 and 1992, biological and chemical agents are of increasing concern. The Centers for Disease Control and Prevention has classified biological terror agents into three categories, based upon their potential to cause morbidity and mortality. Category A agents, such as anthrax, botulism, plague, and smallpox, are classified as high-priority agents because of their ability to inflict high mortality and heavily tax public health and medical resources. Category B agents, such as ricin, typhus, and *Cryptosporidium parvum*,

are moderately easy to disseminate, but would cause lower rates of morbidity and mortality. Lastly, Category C agents, such as Hantaviruses and tick-borne encephalitis viruses, are of the third-highest priority because of their status as emerging pathogens that can be engineered for mass dissemination in the future. In addition to conventional bombs, biological, and chemical agents, Goldman noted that nuclear, economic, and cyber attacks are other potential sources of terror.

> Assessing the terror environment not only allows for monitoring of the sources and routes of exposure, but it also helps to prevent and treat diseases by identifying susceptible and resistant population.
>
> —*Lynn Goldman*

After harmful agents are disseminated, they are transmitted through one of four vectors: water, air, soil, or food (Figure 1.1). Therefore, before agents can exert their dangerous effects, they must be transmitted to humans through inhalation, ingestion, or absorption. The agent, vector, and route of exposure all have a significant impact on the type and severity of the health effects on the exposed population. According to Goldman, assessing the terror environment not only enables monitoring of the sources and routes of exposure, but it also helps to prevent and treat diseases by identifying susceptible and resistant populations. With knowledge of the agent, vector, route of exposure, and expected health effects, exposed populations can be treated at an early stage, subsequently reducing death and disability.

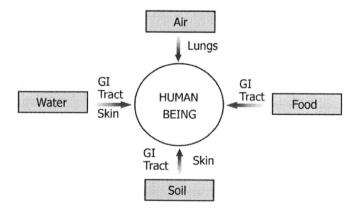

FIGURE 1.1 Possible vectors and routes of exposures by which harmful agents can reach human beings. SOURCE: Ott WR, 1990.

Injury Prevention

In addition to understanding the terror environment, Goldman noted the importance of strengthening both human and physical infrastructure to aid in preventing disasters and reducing their impact, should they occur. To illustrate the benefits of increasing the resistance of structures and people, Goldman discussed how William Haddon's injury prevention matrix can be applied to a terrorist attack. Haddon studied injuries utilizing concepts from engineering, biomechanics, physiology, medicine, and epidemiology. Through his research, he concluded that, like infectious diseases, injuries are the result of an intricate interaction between agents, hosts, and the physical and social environment (Staniland, 2001). While preventing all terror attacks in the first instance would be ideal, it is important that emergency managers and public health professionals plan for success in controlling and limiting the severity of injuries sustained in the event of an attack. The Haddon Matrix's ten strategies, and examples of how each strategy can be used to prevent or mitigate the effects of terrorism, are listed below:

> Building human infrastructure among experts in various fields will help to ensure that trained, experienced professionals are available to respond to future crises.

- *Do not create the hazard.* Prevent terrorism by identifying those who are planning attacks.
- *Reduce the amount of hazard.* If all terrorists cannot be eliminated, reduce their numbers.
- *Prevent release of the agent.* Monitor known terrorists and identify likely threats.
- *Modify release of the agent.* Develop slower-acting explosives.
- *Separate in time or space.* Define a no-vehicle zone near a likely target area.
- *Separate with a physical barrier.* Construct barriers to reduce access to targets.
- *Modify surfaces and basic structures.* Install shatterproof glass in windows.
- *Increase resistance of the structure or person.* Design buildings to withstand bomb forces.
- *First aid and emergency response.* Train greater numbers of volunteers in first aid and rescue skills.
- *Acute care and rehabilitation.* Develop plans and adequate facilities for definitive care (Baker and Runyan, 2002).

To successfully implement injury prevention and control measures, professionals from many diverse fields must work together to prepare for, prevent, and mitigate disasters. Using the injury control matrix developed by Haddon, planners

would consider factors related to the agent (weapon), host (potential victims) and environment (for example, structure of buildings) and whether any of these can be modified pre-event, post-event or during the event. Goldman's application of the Haddon Matrix to terrorism would require successful collaboration between the public health, law enforcement, and medical, engineering, and emergency management/response communities. Goldman further noted that the public health workforce is dominated by professionals reaching retirement age. Therefore, in addition to responding to public health disasters, public health professionals should also invest resources in training new leaders to ensure that they will be ready to work on the front lines of public health as the current workforce retires. Building human infrastructure among experts in various fields will help to ensure that trained, experienced professionals are available to respond to future crises.

EMERGENCY RISK COMMUNICATION

Having well integrated systems of preparedness is only one element in reducing the impact of disasters upon affected individuals and communities. Effective communication before, during, and after disasters, to culturally diverse audiences of wide-ranging scientific literacy, is a critical component of any preparedness effort. According to Dr. Julie Gerberding, Director of the Centers for Disease Control and Prevention, it is essential to communicate with affected communities, the public, the scientific community, and other stakeholders, to provide the information they need to make the best possible decisions concerning their wellbeing within nearly impossible time constraints. The sense of urgency surrounding emergency risk communication distinguishes it from all other forms of health communication, noted Gerberding. Traditional health communication is aimed at pro-

> Effective communication before, during, and after disasters, to culturally diverse audiences of wide-ranging scientific literacy, is a critical component of any preparedness effort.
>
> —*Julie Gerberding*

viding the public with information to promote everyday healthy lifestyles. Emergency risk communication, on the other hand, involves providing information that is, by its nature, incomplete, and likely to change over time. While the emerging risk or hazard may be unforeseen and new to the public health community, the communication must, nonetheless, be science-based. Successful crisis communication can be achieved by skillfully developing messages utilizing tested risk communication theories and techniques, which imply an understanding of human psychology and the needs of people in times of crisis, stated Gerberding. The CDC is starting to pre-test messages before threats occur through the use of focus groups around thematic areas to develop tools for public health.

Further, Goldman echoed the issues raised by Dr. Gerberding and noted that the CDC has organized the schools of public health into a number of centers for public health preparedness that are reaching out to the public, public health officials, and first responders. The purpose is to provide information in advance so that people have the core knowledge and skills to make the process smoother. During the general discussions, some participants reiterated the need for further exploring into research communication. These participants, however, cautioned against reinventing the wheel as it relates to communication. They encouraged research that builds on the 20–30 years of social science studies on risk communication as a necessary part of any research program.

The Emergency Risk Communication Audience

Many individuals operate under the assumption that, even if a disaster does occur, it will not affect them; therefore, when it does happen, few people actually have a plan in place to help them react to the disaster, acknowledged Gerberding. While the disaster, alone, can invoke fear, a lack of adequate resources and complete knowledge of the event can further heighten anxiety and threaten an individual's ability to respond appropriately. During crises, the public looks to politicians, public safety officials, and medical and public health professionals to provide assurance that all possible actions are being taken to alleviate the effects of the disaster, and to recommend actions for individuals to take to ensure their safety.

An individual's emotional response to a crisis is similar to that of any other life-threatening or grave event. Effective emergency risk communications messages reflect an understanding of the different ways that people react in an emergency, and will attempt to manage those stresses in the population. According to Gerberding, the psychological stages of response to crises are:

- Vicarious rehearsal—as a result of continuous news coverage, those located in areas removed from the disaster are still able to participate, vicariously, in a crisis that may not pose any real danger to them. When communicators provide recommended actions to threatened communities, those removed from the danger may also take action, heavily taxing the response effort.
- Denial—denying that the crisis occurred may cause people to delay taking the recommended actions.
- Agitation and confusion—extreme fear and high anxieties may cause people to become agitated or confused by the warnings.
- Doubting the credibility of the threat—emergency risk communication messages may be ignored by those who do not believe that the threat is real, or that it may affect them.
- Stigmatization—following a terror attack, victims believed to be hazard-

ous to associate with (i.e., those contaminated with a biological or chemical agent) may be feared, threatening the social unity within a community.

- Fear and avoidance—this is, perhaps, the most incapacitating of the psychological responses to crises, as fear of perceived or real threats may cause individuals to act irrationally.[2]
- Withdrawal, hopelessness, and helplessness—individuals who do not avoid the threat may, instead, feel powerless to protect themselves from it. This poses a challenge for risk communicators because, as individuals withdraw themselves from situations, their messages may not be heard, or their recommendations may not be acted upon (CDC, 2004a).

When planning crisis and emergency communication messages, in addition to understanding the range of emotions affected individuals may experience, it is important to understand that audiences judge the effectiveness of messages by their timeliness, content, and credibility. According to Gerberding, first messages are lasting messages, as the information provided in the message sets the stage for future communications. The speed of the communication indicates that there is a system in place to respond to the emergency, which can help to ease the public's fear of uncertainty following disasters. In addition, the public is expecting to hear consistent factual information. Inconsistent messages increase anxiety, decrease the likeliness that the public will abide by the communicator's recommendations, and diminish the communicator's credibility for future purposes.

The value of effective risk communication cannot be disputed. During crises, skilled risk communication techniques can provide necessary guidance to audiences of differing ages, educational status, languages, and cultural norms. According to Gerberding, in addition to reaching diverse audiences, messages should be prioritized based on the recipients' distance from and relationship to the threat, as different audiences have distinct concerns. Those closest to the threat should be instructed on how best to protect themselves, while those farther away should be cautioned to remain calm, yet vigilant. The messages will be well received if they are timely, credible, and delivered by a spokesperson that is trusted and familiar with the basic principles of crisis and emergency risk communication.

Emergency Risk Communication Spokesperson

Choosing the appropriate spokesperson to deliver news and recommendations to the public during times of heightened fear and anxiety can be a determin-

---

[2]Quartantelli in 1954 established that panic flight is very rare at any time before, during, and after disaster impact, because the conditions to induce panic are present in only a relatively small number of emergencies. The social solidarity remains strong during most emergency response and few situations occur that can completely break down social bonds (Tierney et al., 2001).

ing factor in the public's response. As Gerberding noted, the spokesperson has four important roles: (1) to remove the psychological barriers within the audience, (2) to penetrate the public's anxiety and gain support for the public health response, (3) to build trust and credibility for the organizations involved in the response effort, and (4) ultimately to reduce the incidence of illness, injury, and death. Through public appearances, the spokesperson gives human form to the organizations charged with the task of resolving the crisis.

> Following the 2001 anthrax attacks, 77 percent of people polled had either a great deal of, or quite a lot of trust in their own doctor to give them advice on how to best protect themselves.

According to a joint study conducted by the Harvard Program on Public Opinion and Health and Social Policy and International Communications Research of Media, PA, immediately following the 2001 anthrax attacks, 77 percent of those polled had a great deal of trust in their own doctor to give them advice on how to best protect themselves. That was followed by high levels of trust in a fire department official, police department official, local hospital official, health department leader, governor, and, finally, a religious leader. On a national scale, 48 percent of those polled had a great deal of trust in the CDC director, followed by: the Surgeon General, the American Medical Association president, the Secretary of the Department of Health and Human Services, the Secretary of the Department of Homeland Security, and lastly, the Director of the Federal Bureau of Investigation (Pollard, 2003).

While the foregoing study clearly indicates that clinicians play an integral role in emergency risk communication, according to Gerberding, regardless of the communicator's professional background, there are five key rules that all spokespersons must follow to increase the likelihood of a successful communication. First, to provide a greater chance that the message will be acted upon, the communicator must exhibit sincere empathy for those affected by the disaster. Risk communication experts at the Centers for Disease Control and Prevention estimate that 50 percent of a spokesperson's effectiveness directly relates to their capacity to communicate that they genuinely care about what is happening (Figure 1.2).

Second, Gerberding noted that with continuous news coverage, it is impossible for the spokesperson's message to always be the first. As explained above, first messages indicate that the responding agencies are prepared and competent to deal with the crisis. To aid in accelerating the network of communication, it is critical to have a command center and emergency communications system, where members from all responding agencies can communicate so that the appropriate information is disseminated to the public.

Third, the content of the risk communication message must be accurate and consistent with other messages. Being wrong not only decreases the public's

Listening/Caring/
Empathy
50%

Competence/
Expertise
15-20%

Dedication/
Commitment
15-20%

Honesty/
Openness
15-20%

FIGURE 1.2 The spokesperson's ability to embody empathy and caring is the single most important factor in gaining the audience's trust. SOURCE: adapted from Covello VT, 2001. Reprinted with permission.

confidence in the response effort, it also destroys the credibility of the spokesperson's organization for future communications.

The fourth rule is that the spokesperson must be honest. According to Lynn Goldman, of The Johns Hopkins University Bloomberg School of Public Health, research has shown that, if the public is given honest information, inappropriate behavior will be less likely and many people may even be comforted by the message. In addition, Gerberding noted the value of refraining from delivering completely negative messages. As a result of the emotional component of disasters, if the spokesperson needs to deliver one negative message, it should be balanced with at least three positive messages. Negative words are very difficult to overcome in the context of a crisis; therefore, honest messages should be delivered using positive or neutral words. At the same time, Gerberding emphasized the value of not over-reassuring the public because, if the crisis situation intensifies, the spokesperson and the organization will lose their credibility. Instead, the communicator should acknowledge the uncertainty surrounding the disaster, express that a process is in place to learn more about it, acknowledge the public's fear and misery, and ask that the public work with responders to find a solution (CDC, 2004a).

> As a result of the emotional component of disasters, if the spokesperson delivers one negative message, it must be balanced with at least three positive messages.
>
> —*Julie Gerberding*

Finally, according to Gerberding, the fifth risk communication rule is to get help. If information is unknown, the spokesperson should tell the public that, but, at the same time, emphasize that everything possible is being done to find the answer. Those five rules are useful in creating and communicating an effective risk message; however, the actions suggested in the message will not be acted upon unless the message is disseminated to the public using appropriate methods of delivery.

Working with the Media to Communicate Risk

The media is the fastest, and, in some cases, the only means to circulate important public health information to the public during a crisis; therefore, working with the media is critical to successful communication. While the media is expedient as an emergency broadcast system, members of the media may not have the background knowledge to immediately understand the scientific or technical issues surrounding many disasters. Thus, it is important for spokespersons to speak plainly in order to avoid miscommunication and misinformation. Furthermore, prior to issuing a press release or a statement to the media, Gerberding suggests anticipating and preparing responses to potential questions to ensure that appropriate answers are provided to help achieve a positive health impact.

To disseminate information to the public in the event of a loss in electrical power, Lynn Goldman emphasized the importance of crank radios, battery-powered radios, and landline telephones. Unfortunately, many American homes and businesses do not have such essential preparedness equipment, which can result in a complete breakdown in communication during disasters. It is, therefore, the role of the Centers for Disease Control and Prevention and other agencies and NGOs to communicate the value of preparedness before the next disaster occurs.

Emergency Risk Communication at the
Centers for Disease Control and Prevention

According to Gerberding, emergency risk communication at the CDC is a science application that has rapidly developed since the 2001 anthrax attacks and has been strengthened during the recent SARS, West Nile, monkey pox, avian flu, and influenza outbreaks. Through its Futures Initiative, the CDC has a new capacity to help individuals, stakeholders, and communities obtain the information they need to make the best possible decisions about their well-being. To achieve its goal, the CDC has established a global communications command center with the ability to videoconference with the Department of Health and Human Services, the Department of Defense, the State Department, the Food and Drug Administration, the National Institutes of Health, and the World Health

Organization. This helps to ensure that the CDC always has the latest information available to distribute to the public.

The CDC has an emergency communications team made up of expert communicators who translate scientists' findings and recommendations to the media, laboratories, clinicians, state and local departments of health, academia, national and international corporations, and other stakeholders in public health crises (Figure 1.3). The team is currently using focus groups to pretest messages before the threat occurs, so they will have the necessary tools available to broadcast information to the public at a moment's notice.

Recognizing clinicians' vital role in emergency risk communication, the CDC has employed a tiered approach utilizing health educators, clinical specialists, and frontline clinicians to develop two different types of clinician communication (Figure 1.4). First, the "Just in Case" communication trains clinicians in anticipation that a public health crisis might occur. Second, the "Just in Time" communication gives clinicians the latest information to help diagnose, treat, and communicate with patients during a crisis. According to Gerberding, "It is clear to us as an agency that the ability for scientists to translate their science to the media, to the public, and to other communities is central to our success."

To aid organizations in developing and disseminating emergency risk communication messages to the public, the CDC has created a detailed website, which can be accessed at: http://www.cdc.gov/communication/emergency/erc_overview.htm.

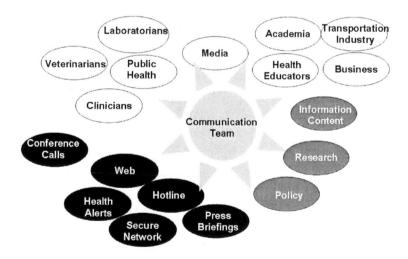

FIGURE 1.3 The CDC's emergency communications team translates scientists' findings and recommendations to the media, laboratories, clinicians, state and local departments of health, academia, national and international corporations, and other stakeholders in public health crises. SOURCE: CDC, 2002. Reprinted with permission.

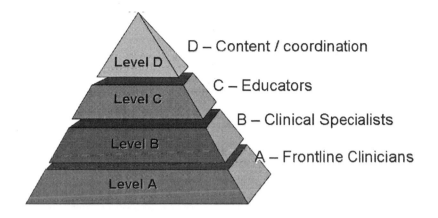

FIGURE 1.4 The CDC uses a tiered approach of health educators and clinical specialists to distribute information to clinicians both prior to, and during, public health emergencies. SOURCE: CDC, 2002. Reprinted with permission.

2

Linking Hazards and Public Health:
Case Studies in Diasters[1]

Using three different types of disasters, heat waves, earthquakes, and complex humanitarian crises, the speakers and Roundtable members explored the complexity of responding to health needs, the public policy underlying the response, and the short- and long-term health-related needs. Speakers discussed examples of planning and response to specific events in the United States such as the Chicago heat wave of 1995 and the Loma Prieta earthquake of 1989, as well as events occurring internationally. Discussion continued regarding how communities at all levels, from small towns and counties to major cities, may link major disasters to public health, strategies for creating a plan of action, and implementation of these programs through managing disasters as they evolve.

SOCIAL AND HEALTH EFFECTS OF A DISASTER—
THE HEAT WAVE

Heat waves are often the "overlooked" natural disaster, frequently not recognized by the media or government as a public health risk and omitted from the disaster literature. Eric Klinenberg of New York University stated that one of the immediate challenges faced by individuals attempting to persuade the public and government officials of the direct health threat presented by heat waves is the lack of imagery associated with a heat wave. Often people conjure images of children playing near a fire hydrant or people sipping lemonade on a front porch as ways to skip the heat. Compare this to the image of cars twisted around trees or houses carried off their foundations down a river. While heat waves do not have the same destructive properties as a tornado or an earthquake, more lives

---

[1]This chapter was prepared by Victoria Blaho from the transcript of the meeting. The discussions were edited and organized around major themes to provide a more readable summary and to eliminate duplication of topics.

More lives have been claimed in the United States over the past fifteen years by heat than by all other extreme weather events—lightning, tornadoes, floods, and hurricanes combined.

—Centers for Disease Control and Prevention

have been claimed in the United States over the past fifteen years by heat than by all other extreme weather events—lightning, tornadoes, floods, and hurricanes—combined (CDC, 2002). He cited an editorial in the *New England Journal of Medicine*, published just a few months after the Chicago disaster, which states that compared to other kinds of disasters that kill far fewer people, the heat wave in Chicago was forgotten almost as soon as the weather changed (Kellermann and Todd, 1996).

The Furnace—The Dynamics of a Heat Wave

More heat-related deaths occur in cities than in rural areas because stored heat dissipates slower in urban areas. This is due to the density of brick and stone buildings, paved streets, and tar roofs that store heat and radiate it like a slowly burning furnace to create a "heat island." The worst heat disasters, in terms of loss of life, occur in large cities when a combination of four factors occurs for a period of several days:

- high daytime temperatures
- high humidity
- warm nighttime temperatures which prevent dissipation of stored daytime heat
- abundance of sunshine, which can increase the heat index by 15°F

Examined independently these atmospheric conditions may be of little consequence. However, in combination they can create an urban environment where infrastructure stores heat and continually releases it throughout the night until the next day, when more heat will be absorbed for the cycle to continue until temperatures drop.

The Role of Socioeconomic Factors

Socioeconomic problems are risk factors for susceptibility to heat-related illness. Klinenberg pointed out that understanding the relationship between neighborhood conditions and vulnerability can help cities target their responses to those areas with populations that may be hit the hardest. For example, lower-income individuals may not have air conditioning or may hesitate to turn it on due to cost. Often they live in high-crime areas and may be afraid to open the

windows, creating an indoor environment equivalent to a greenhouse with little air circulation and increasing temperatures. Additionally, the mentally ill, who are more likely to be alone because of difficulty in gaining and maintaining social support, may also have difficulty cooling down or avoiding severe sunburns due to their medications. The risks of social isolation incurred by groups such as the elderly and mentally ill are only compounded as neighborhoods evolve and the cultural, ethnic, and linguistic composition of the community changes.

A Public Health Policy Example: Recent Heat Waves

Within the past decade there have been two heat waves with catastrophic results: the Chicago heat wave of 1995 and the European heat wave of 2003. In July of 1995, while the entire Midwestern United States endured an abnormal increase in average temperature, the urban heat island of the Chicago area experienced the highest temperatures recorded since measurements began at Midway Airport in 1928, with daytime temperatures peaking at 106°F. Over 700 people in the city died during this heat wave that lasted about five days.

In the summer of 2003 over 35,000 people died in Europe when an abnormal weather system that lasted for about three weeks aggravated the conditions that had been set in motion by an early and unusually warm spring and low rainfall (Rasool et al., 2004). While the death rate in Chicago from 1995 was actually about identical to the death rate in France, the deaths in Chicago came after two or three days of elevated heat and humidity, whereas the deaths in Europe occured toward the end of the heat wave. Although there is some debate, the delayed deaths in Europe may have occurred due to the temperatures and unusual atmospheric conditions reaching a critical juncture, whereas the conditions in Chicago resulted from a sudden onset of extraordinary atmospheric conditions.

Complicating Factors in the Chicago Heat Wave of 1995

While high temperatures and unusual weather conditions are essential components of a heat wave, there are often a number of compounding issues, such as a lack of communication between government officials and a failure of critical infrastructure. With no official disaster plan in place to address the crisis, the effect of the heat in Chicago was compounded by a number of complicating factors:

- Communications: During a 5-day period in which not only were there 700 more deaths than during a normal period, but also thousands more were hospitalized, some paramedics who first arrived on the scene reported that their own departments refused to release additional ambulances and staff to cope with the workload.

- Power failure: When ConEd power failures knocked out fans and air conditioners that summer, 49,000 households were affected, and hundreds of Chicagoans died (Careless, 2004).
- Inadequate facilities: City officials did not release an emergency heat warning until July 15, the last day of the heat wave. Because of the delay in issuing an excessive heat advisory, emergency measures such as Chicago's five cooling centers were not fully utilized, severely taxing the medical system as thousands were taken to local hospitals with heat-related problems.
- Lack of understanding: Klinenberg stated that a large part of the problem was with reporting and recognition—a failure and in some cases a refusal to recognize the extent of the damage and the potential for further risks, such as the difficulty experienced by paramedics attempting to convince high-ranking officials that the health problems they were addressing constituted a disturbing trend and a serious threat.

The "Social Autopsy" of the 1995 Chicago Heat Wave

In situations such as the aftermath of the Chicago heat wave of 1995, Klinenberg proposes that the affected community analyze the response and results of a disaster immediately after it occurs, which Klinenberg called social autopsy. He expressed the idea that this is especially important because excessive heat disasters are one of the few disasters where deaths are preventable, and also suggests that being relatively open with the results of such an evaluation will allow community leaders to generate a general understanding about public health.

The Effects of Chicago Autopsy Results

While the Chicago heat wave was a terrible disaster, Klinenberg noted that this event could be thought of as leading to a checklist for the new Chicago disaster response plan. Various groups such as the Department on Aging now compile lists of elderly people who live alone and might need assistance so workers can call or visit those residents to alert them that a dangerous weather system is on the way. Similarly, the city now also opens up a heat line for updated safety information. Another crucial step was the implementation of a monitoring system for emergency room admissions and the activity of paramedics, as the danger posed by a weather system can immediately be understood by monitoring the health impacts reported by front line responders.

Bringing New Life to Disaster Response

According to Klinenberg, before a heat wave has arrived a city should examine its infrastructure such as water systems and communications to deter-

mine how, when and where response systems will be needed and how to make them easily accessible and moderately simple to implement. Another critical question that cities need to address is deciding at what point in the slow onset of an event such as a heat wave it must be addressed as an imminent threat. The corollary to this is how to acknowledge and publicize the imminent threat without creating undue public alarm. Once a heat wave is declared a public health emergency, the response plan can be implemented. Klinenberg also emphasized the need to perform a social autopsy after the disaster, when the protocols have been carried out and the damage has been tabulated, to dissect the response and provide immediate feedback.

Applying Lessons Learned to Other Disasters

During a time of crisis, it may be useful to use the lesson learned from one disaster for responding to other disasters. For example, a participant who is a staff member from the Office of Emergency Management for New York City noted that following the 1999 Chicago heat wave, the city of New York created a network to address the special needs of the elderly and the particularly vulnerable in the event that a similar event happened in New York City. The at-risk individuals were identified because they were receiving home-based care or nursing services, having contact with the Department for the Aging, or participating in activities at senior citizen centers. While this network was designed to respond to potential heat waves, its was fully tested following the World Trade Center (WTC) attacks. When implemented on September 11, the Office of Emergency Management needed to contact the 3500 individuals within the affected area to ensure that the individuals were receiving care. Within 2 hours of the disaster, a call center was initiated, and within 24 hours, all but 30 individuals were contacted. A joint team of Red Cross volunteers and construction workers began to locate the remaining individuals because the EMS and the police were occupied with events at the WTC site. The participant noted that the use of the call center was successful; however, as this example illustrates, there needs to be more flexibility in the planning as events unfold. For example, the use of able bodied volunteers to check on vulnerable populations while emergency personnel are busy with the crisis.

THE ROLE OF INFRASTRUCTURE DURING A DISASTER

Infrastructure and public health are not necessarily thought of as interconnected areas, although their relationship to one another can be profound, especially in disasters. Part of the traditional purpose of infrastructure is to protect human health, and so disaster planning needs to be explicitly incorporated into infrastructure design, operations, and maintenance. In addition, infrastructure services are interrelated, which has implications for public health. Discussions of

infrastructure tend to dissect and analyze infrastructure sectors individually. The impact of transportation, utilities, and communications upon each other, however, should be examined as a whole, since these services serve and impact the same customers, noted Rae Zimmerman of New York University's Wagner Graduate School of Public Service.

Infrastructure Under Non-Disaster Conditions

In urban areas the quality of the built environment, which includes infra structure, dramatically affects the health status of all urban residents. Zimmerman stated that air and water quality, for example, encompasses infrastructure-related public health issues for all urban residents on a day-to-day basis:

- Under non-disaster conditions transportation is the single largest contributor to emissions of several air quality pollutants, and similarly, the generation of electric power is a key contributor to several air quality pollutant emissions (Wright, 2005: 580, based on U.S. EPA information).
- There are an estimated 1.3 million cases of water-related disease in the United States (quoted in Zimmerman, 2004: 86, citing Water Infrastructure Network, 2000: 1-2 and Payment et al., 1997). Infrastructure-related factors associated with these diseases can include poorly-planned population expansion that is not accompanied by increased water and wastewater treatment capacity or defects in engineered systems such as water treatment and distribution or waste water management systems.

These are just two examples of the multitude of public health issues that relate to infrastructure.

Infrastructure in the Short and Long Term

Infrastructure decisions rarely reflect an emphasis on public health, and people making those decisions often have little training in public health areas. In addition, regulators and planners in infrastructure areas often do not coordinate with public health professionals, and thus, all of the short- and long-term effects of the interaction between environment, infrastructure, and health may not be considered.

The short-term effects of these decisions are that in times of disaster, structural damages may not be avoided as effectively as they might otherwise be, such as the collapse of freeways or buildings in an earthquake, and the long-term effects and their impact on health are often not tracked. In the area of infrastructure design this can be demonstrated by examining the collapse of the Cypress and Embarcadero Freeways during the Loma Prieta earthquake of 1989. These two double-decker freeways suffered severe structural damage (Figure 2.1). Sev-

FIGURE 2.1 The Cypress Street Viaduct (left) before the earthquake of 1989 destroyed much of the structure (right), requiring its demolition. SOURCE: EBMUD Seismic Improvement Program. Available [on-line] at: http://home.pacbell.net/hywaymn/Cypress_Viaduct_Freeway.html

eral thousand people were injured and dozens were killed in that earthquake (Tubbesing, 1994). Zimmerman pointed out that accounts of the collapse indicated that the short-term decisions made about construction of infrastructure and later retrofitting only one side of the freeway may have contributed to instability.

Government and industry tend to focus on immediate rather than long-term impacts, e.g., giving greater emphasis to the effects and cost of initial construction under normal conditions of use rather than to structural stability in the event of a hypothetical disaster in the future. Policy is designed accordingly and thus may not properly identify or address many important areas related to disasters. Effective environmental and public health regulations for infrastructure must involve collaboration between all involved parties, proper analysis of short- and long-term environmental and health impacts, and the development and implementation of effective policies that respond accordingly, concluded Zimmerman.

Infrastructure Organization and Management

The manner in which infrastructure is organized and managed can have a direct impact upon the vulnerability of a society in times of disaster. The dramatic centralization of virtually all areas of infrastructure has become a conscious policy for economic and managerial reasons. Figure 2.2 illustrates this concept of centralization. As communities move from a small population density to a much higher population density, there is, for example, an evolution in the provision of water supply services, from wells to community water supply systems to urban water treatment plants. Similarly, the natural evolution of wastewater treat-

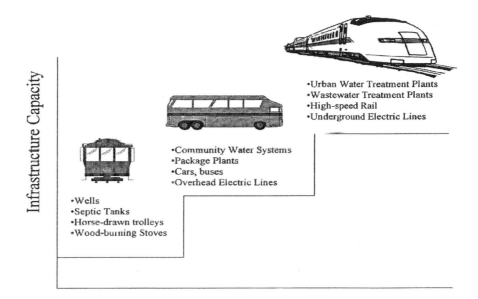

Population/Population Density

FIGURE 2.2 Our society has been evolving toward ever more centralized systems for transportation, water, electricity, and other utilities. SOURCE: Rae Zimmerman, unpublished. Reprinted with permission.

ment is a movement from septic tanks to package plants to large wastewater treatment plants, and transportation has grown from horse-drawn trolleys, to cars and buses, and more centrally controlled or managed large high speed trains and airplanes. With electric power, individual stoves have given way to electric heating capacity provided via overhead electric power lines followed by underground lines, and from smaller electric power plants to larger ones based on energy sources such as coal or nuclear power, noted Zimmerman.

Society is also becoming increasingly reliant upon infrastructure networks that often span large distances, stated Zimmerman, noting that the United States has almost 4 million miles of highway, 10,000 miles of track for city and regional rail, 22,000 miles of track for long distance passenger travel, and 170,000 miles of freight railroads (summarized in National Research Council, 2002). Beyond transportation, the United States boasts close to 1 million miles of water supply line and similar numbers for wastewater piping, providing a convoluted set of networks vulnerable to natural and terrorist threats.

In addition to vulnerabilities created by the extensive network of distribution systems, interdependencies found among the separate components of infrastructure can also potentially create vulnerabilities due to cascading and escalating effects. The individual units of infrastructure are each vulnerable independently to physical and electronic disruptions, and a dysfunction in one can have severe consequences in the others. While one aspect of a region's or the nation's infrastructure may be more sensitive to a disruption, they are all dependent upon one another to varying degrees.

> The individual units of infrastructure are each vulnerable independently to physical and electronic disruptions, and a dysfunction in one can have severe consequences in the others.
>
> —*Rae Zimmerman*

Infrastructure: Choices and Trade-Offs

Difficult choices often have to be made regarding risks and benefits when considering the effects of infrastructure options upon the health of a population, such as the use of diesel fuel for emergency power back-up to generators versus the health effects that may result from diesel fuel emissions. Decisions made by government and industry must involve the decoupling of infrastructure and shift our dependency from centralized energy sources to renewable energy sources such as solar, waste, wind, and other relatively newer technologies that can operate in a decentralized manner, stated Zimmerman. This will ensure that the infrastructure systems crucial for the day-to-day functioning of our communities and the nation can withstand a disaster and maintain the trust of the public.

COMPLEX DISASTERS AND PUBLIC HEALTH

There isn't a single internationally adopted definition of what constitutes a complex disaster, stated Jean-Luc Poncelet, of the Pan American Health Organization (PAHO). A complex emergency is a term primarily used by the United Nations (UN) that refers to a crisis that overwhelms nations due to civil disturbances, war, deep political crisis, etc. Due to the chaos, the entire nation becomes dysfunctional and humanitarian intervention from a foreign source is needed.

The health risks associated with complex emergencies tend to be very poorly documented and often biased. This is because the majority of the morbidity and mortality information is provided by non-national, well-meaning individuals, but often with little or no knowledge of the origin of the conflict, the traditions and culture, or even the language. This can result in the production of copious amounts of data that are only partially analyzed and are frequently distorted, observed Poncelet. Thus, PAHO and the World Health Organization (WHO) as

part of their mission focuses on strengthening the national capacity to respond to crises. Even though this may be difficult because in a complex emergency there is no government, the fate of the government is at stake, or the authority itself is part of the conflict.

The Risks of International Assistance

An increasing number of people and organizations are intervening in the humanitarian field. On the one hand, this is beneficial because it brings attention to the situation that is occurring. On the other hand, the complications involved in attempting to communicate with and coordinate the efforts of large numbers of international organizations can become more of a burden than a blessing, noted Poncelet. All of them attempt to assist countries in their specific field, so efforts can become extremely complicated, especially if there is a strong political or media influence, as is commonly present in a complex emergency.

The weakening of remaining local response capacity by setting up parallel coordinating mechanisms is perhaps the greatest risk of international assistance. The risk is that organizations and individuals believe that just because there are good intentions, beneficial short-term results, and excellent specialists this automatically means that they are going to do a good job, observed Poncelet. The involvement of the local and usually fragmented network is critical to attaining any level of success. International helpers can actually become a burden on the national system. This would occur in a situation such as the deployment of expensive field hospitals that arrive late and then remain in the country after stabilization with high maintenance costs. Poncelet posed the question of why send 300 people for one week who don't speak the language and don't know the context, to assist in a complex emergency; a situation that has happened repeatedly. Unfortunately, common sense is often abandoned in the rush for visible action to satisfy the international public, and not the local needs. The money used for a large scale operation could have been used toward people and supplies in smaller quantity for a longer period of time. It would have been less visible internationally, but more effective locally.

Direct and Indirect Impacts

The direct impact of these conflicts on public health is usually measured by the mortality, noted Poncelet. For example, in Bosnia, the mortality of traumas in 1992 increased dramatically within the time-span of a few months, from 22 percent mortality of trauma cases to 78 percent mortality, an increase that was directly linked to the civil war and international intervention.

Hospitals and the Red Cross Society emblems were previously considered to be safe harbors even during a complex emergency; however, over the last 5–10 years there has been an increasing trend towards targeting the health

services themselves as a war tactic. For example, the killing of victims in ambulances, hospitals staff being given instructions not to attend to parts of the population by the leadership of a guerilla movement, and massacres taking place in hospitals have become more commonplace.

Indirect impacts do not provide stark images. They are silent, but the most serious ones to be attended by humanitarian health professionals. They vary in origin, including:

- long-term interruption of health services due to impairments in access or security
- the need for provisional housing
- interruption of infrastructure, such as water systems, electricity, transportation
- general insecurity and psychological impact of events on population and staff
- limited access to food due to lack of income, lack of adequate stores, destruction of crops, etc.

Poncelet observed that the disruption of basic needs and supplies can often prove far more devastating than the direct impact of the disaster itself. For example, the major issue for victims of the 1996 earthquake in El Salvador was not a lack of physicians or medications. Rather, the lack of access to water for drinking, sterilization in hospital procedures, cooking, and cleaning proved to be the most disruptive to citizens.

> We prefer to see victims attended by hospital physicians, but what will save the largest number of lives is the fixing of the pipelines that will ensure the functioning of the kitchen, the laundry, and the sterilizers.
>
> —*Jean-Luc Poncelet*

Public Health Needs

It is important to divide the needs of individuals suffering during and after a complex emergency into two categories: short-term and long-term needs. Short-term refers to the more immediate assistance, an area of specialty better attended by humanitarian professionals. These professionals are trained to address the most urgent requirements of victims, such as food, water, shelter, sanitation, medication, epidemiological surveillance, and logistics. The specialists dealing with long-term issues are from a completely different professional background. Poncelet pointed to the situation in Angola. After years of civil war, the system in place is dealing with issues that cannot be only attended to by humanitarian specialists. Most of these must be dealt with by long-term specialists such as

developers and planners. These professionals have the knowledge to deal with chronic issues in complex situations such as the implementation of functioning health care programs with local resources. The setting up of the response to a crisis is the business of humanitarian professionals and the running of the extended crisis must be dealt with by developer specialists.

Challenges

The division of labor between the different specialties is something that deserves more attention than it commonly receives, stated Poncelet, presenting a challenge for international aid organizations. The nature of a complex emergency is that it is usually a long-lasting event, with no possibility of being resolved in a few months, like in the case of the aftermath of a tornado or a flood. Humanitarian specialists are the most visible respondents but must also stay true to their area of expertise. Poncelet also noted that in order for the response to be most effective, both groups must be present as soon as possible, meaning that groups who are working on the long-term challenges should also be present during the early stages of the response and integrated into the work of the humanitarian specialists in order to coordinate their efforts to provide a long-term solution and not just a momentary lift.

Effect on Mental Health

The provision of mental health services has traditionally been overlooked as a priority need in the case of complex disasters. However, this view is changing with an accumulation of research on long-term coping and functioning skills of affected populations, noted Poncelet. Depression and post-traumatic stress disorder (PTSD) are common disorders in war-torn regions and soldiers returning from wars. The impact of wars on mental health can linger for years after the war ends, with affected populations having a lower level of social functioning than non-affected populations. Poncelet asserted that more emphasis should be placed on long-term effects, such as a reduction in resilience due to the absence of a structured environment, schools, and family.

> Mental health has traditionally been overlooked as a priority need for providing assistance in the case of complex disasters.
>
> —*Jean-Luc Poncelet*

NGOs and Complex Disasters: Challenges

The mission of PAHO/ WHO is to face natural disasters and complex emergencies as part of the inter-America system and as a regional office of the World

Health Organization. The goal of the organization, as stated by Poncelet, is to work with local authorities primarily before disasters in the areas of prevention, mitigation, and preparedness, but also to aid in the response to disasters, based on the local response capacity.

PAHO as a Model of Preparedness and Response

PAHO/WHO, in coordination with other governmental and non-governmental organizations, views preparation as its best investment. This includes such activities as:

- helping to implement and continuously strengthen national disaster programs
- training health sector personnel
- inter-institutional and inter-sectoral coordination mechanisms

The level of the response from PAHO/WHO is dependent first upon the local response to a disaster and then, to a lesser degree, the extent of the international response.

All aspects of potential needs cannot be prepared for in every locality, admitted Poncelet; therefore, PAHO also devotes some of its energies to regional response mechanisms. A great deal of this effort is devoted to the coordination of international health assistance. This allows for technical cooperation between local and international officials with an independent assessment of specific needs in the current situation and the ability to mobilize international resources to complement the local and national response, if necessary.

If You Don't Know, Don't Go

In summary, Poncelet put forth three major points regarding complex disasters and their impact on public health: the need for quality information, protection of public health services, and availability of appropriate and timely expertise.

In most complex disasters only fragmented information is available, and this information must be viewed with caution because of the risk of potential bias in reporting. As discussed above, to avoid this risk it is necessary to have studies done locally with local users, asserted Poncelet. Also important is the protection and utilization of the existing public health services, even if they are not in the conflict areas.

The final point is evaluating the appropriateness of expertise. As mentioned previously, it is important to delegate responsibility and ensure that people who are in charge of the humanitarian help and quick response act only in the realm of their expertise and at the same time have them working much more closely

with developers. These two groups of people working together are the only alternative to assist countries in crisis.

Poncelet suggested that the best method of alleviating some of the strain on public health posed by a complex emergency is to help the remaining local structure to respond and prepare for a crisis as poorly coordinated international intervention can prove more dangerous than effective.

3

Preparedness and Response:
Systems, Supplies, Staff, and Space[1]

As a result of the unpredictability and increasing frequency of both natural and manmade disasters, medical and public health systems throughout the United States often find their resources taxed beyond their capabilities. While catastrophic events occur locally, placing immediate importance upon local resources and preparedness, according to William Raub, of the U.S. Department of Health and Human Services, preparedness and response must be multifaceted. First, it requires a vertical integration of local, state, and federal government resources. While state and federal assets are not immediately available to local responders, within 4 to 24 hours they can be mobilized and greatly enhance the capabilities of the response to an event of any nature. Preparedness and response are principally government roles; therefore, federal, state, and local elected officials must collaborate to better understand the potential risks of disasters and how to best protect society from them. Second, preparedness requires horizontal integration between public health, health care, veterinary, agricultural, emergency management, and private sector assets to strengthen the response infrastructure at each level.

When the current gaps in public health and health care are considered in the context of an incident involving a weapon of mass destruction (WMD), preparedness and response capabilities take on even greater importance. Tactical nuclear weapons, possibly obtainable in Western Europe, could destroy much of the human and physical infrastructure relied upon for a response effort; therefore, for local responders to provide even a minimal level of care for mass casualties, federal and state governments must provide supplemental assets. While the United States is clearly vulnerable to such an attack, some officials, not understanding the seriousness of the threat, do not believe that the risk

---

[1]This chapter was prepared by Melissa Cole from the transcript of the meeting. The discussions were edited and organized around major themes to provide a more readable summary and to eliminate duplication of topics.

warrants the trade-offs necessary to address it. According to Raub, three key disagreements exist among officials, (1) the likelihood of a terrorist attack that will result in such mass casualties, (2) the balance of investment between the general enhancement of public health infrastructure and the special emergency response capabilities needed to respond to an event of such magnitude, and (3) the necessary balance of investment among local, state, and federal government assets. To best protect the public's health, Raub noted the need for better communications concerning the nature of the risks and the vulnerabilities and trade-offs in addressing them, as well as vertical and horizontal integration of assets to strengthen the ability of the United States to respond to large-scale events.

To determine the local, state and federal resources that are necessary to respond to disasters, Jonathan L. Burstein has suggested a model defining the preparedness and response problem in terms of systems, supplies, staff, and space (Burstein, 2004). The systems component of the model seeks to address the communications and logistics needed to prepare for and respond to crises. The supply variable addresses the drugs, vaccines, and basic necessities—housing, food, and water—that victims need, and how to best distribute those resources among affected communities. Staff considerations include training and credentialing adequate numbers of volunteers and ensuring their safety throughout the response effort. The final component of the model, space, takes into account the physical space needed for patient care, isolation, if necessary, and the distribution of community prophylaxis. Upgrading the public health and health care systems by strengthening systems, supplies, staff, and space, will allow local, state, and federal governments to better respond to disasters.

SYSTEMS

During recent disasters in the United States, responders have encountered numerous problems, including confusion over the jurisdiction responsible for coordinating the response effort; an inability to communicate the vulnerabilities and risks before, during, and after the crisis; difficulties in getting responders to the disaster site while moving victims away from it; and problems distributing essential resources among those who need it most. To alleviate those problems during future responses, the U.S. Department of Health and Human Services has made improvements in state and local preparedness by providing funding and guidelines for all 50 states, the District of Columbia, the territories, and three major urban areas—New York City, Chicago, and Los Angeles County. According to Raub, the Department hopes to improve the response capabilities for bioterrorism and other disasters, while overcoming decades of neglect in the public health infrastructure with respect to containing infectious disease outbreaks.

Funding Preparedness Efforts through Cooperative Agreements

The cooperative agreement is the funding instrument utilized by the Department of Health and Human Services (DHHS). Recognizing the importance of integrating the health care system response plans with the public health department plans, DHHS has incorporated both hospital and public health preparedness standards into the cooperative agreements. To obtain funding, jurisdictions and hospitals must demonstrate, through their proposals, a willingness to collaborate in planning an effective response. As Raub noted, since fiscal year (FY) 2002, DHHS has spent over $2.7 billion on public health preparedness efforts through cooperative agreements administered by the Centers for Disease Control and Prevention (CDC), and $1.1 billion on hospital preparedness cooperative agreements, administered by the Health Resources and Services Administration (HRSA).

Similar to grants, cooperative agreements provide hospitals, states, territories, and cities with structured "critical benchmarks," or standards, which must be met using the funding given to them. DHHS uses these benchmarks as important indicators of progress and recognizes that, while attaining any one of the standards does not guarantee preparedness, failure to achieve any of them is a certain indicator that the hospital or jurisdiction is inadequately prepared to respond to bioterrorism or other health emergencies. The guidance provided by DHHS has encouraged states, territories, and cities to make improvements in seven key areas: preparedness planning and readiness assessment, surveillance and epidemiology, laboratory capacity for handling biologic agents, laboratory capacity for handling chemical agents, health alert network and information technology, communicating health risks and health information dissemination, and education and training (DHHS, 2004a). It is essential that jurisdictions work with their hospitals to ensure preparedness in those seven areas. With their HRSA cooperative agreements, hospitals are to focus on six areas: governance, regional surge capacity to treat victims, emergency medical services, hospital linkages to public health departments, education and preparedness training, and terrorism preparedness exercises. Interspersed throughout the hospital and public health focus areas are activities related to smallpox preparedness (DHHS, 2004a).

> While attaining any one of the critical benchmarks does not guarantee preparedness, failure to achieve any of them is a certain indicator that a hospital or jurisdiction is inadequately prepared to respond to bioterrorism or other health emergencies.
>
> —William Raub

Considering the broad nature of the focus areas, the Department of Health and Human Services has developed 25 critical benchmarks for the (FY) 2004 CDC administered cooperative agreements. While the Department views the achievement of each benchmark as a building block for future preparedness

milestones, Raub discussed four priority standards for jurisdictions and hospitals to accomplish:

- Develop or enhance plans that support local, statewide, and regional responses to bioterrorism and other public health threats and emergencies. Plans must demonstrate the jurisdiction and hospital's ability to rapidly administer vaccines and other pharmaceuticals and to perform healthcare facility based triage. Hospitals should be included in the development of emergency mutual aid agreements in the event of a disaster.
- Develop and maintain a system to receive and evaluate urgent disease reports and to communicate with and respond to the clinical or laboratory reporter on a 24/7 basis.
- Complete and implement an integrated response plan that directs public health, hospital-based, food testing, veterinary, and environmental testing laboratories in responding to a bioterrorism incident.
- Implement a plan for connectivity of key stakeholders involved in a public health detection and response.

As Raub pointed out, during the 2003 Severe Acute Respiratory Syndrome (SARS) epidemic, 21st century information technology converged with 19th century public health and medical practices. Other than movement restriction, isolation, and other containment methods, the United States public health and medical systems lacked means to protect the public's health, e.g., no SARS-specific diagnostics, therapeutics, or vaccine were available. With the implementation of the above critical benchmarks, improved surveillance, epidemiology, reporting, and health communication will enable public health officials to detect outbreaks earlier and ensure that warnings and recommendations are disseminated to all Americans in a timely manner.

> During the 2003 Severe Acute Respiratory Syndrome (SARS) epidemic, 21st century information technology converged with 19th century public health and medical practices.
>
> —*William Raub*

The National Response Plan

While local jurisdictions provide the initial response assets needed to respond to crises, complex emergencies will require help from federal and private-sector resources; therefore, a single, unified, comprehensive national effort is necessary to upgrade the United States' readiness system, with the ultimate goal of increasing the nation's preparedness and response plans, stated Lew Stringer, U.S. Department of Homeland Security. On February 28, 2003, President George W. Bush issued Homeland Security Presidential Directive 5 (HSPD-5), ordering the

development of a National Response Plan (NRP) under the direction of the Secretary of Homeland Security, to ". . . integrate Federal Government domestic prevention, preparedness, response, and recovery plans into one all-discipline, all-hazards plan" (U.S. Executive Office, 2003).

Under the NRP, a standardized model of emergency management procedures, called the National Incident Management System (NIMS), will be created to ensure that all federal departments and agencies, state and local authorities, and private and non-governmental entities partnering with the federal government can unify and synchronize their efforts to prepare for, respond to, and recover from any type of disaster or security concern. While recognizing that each incident is unique, the all-hazards plan will be applied to natural disasters, power outages, chemical spills, civil or political incidents, and designated special events, such as the Olympics and the State of the Union address (DHS, 2003). However, a few participants noted that while the NRP has been issued, it hadn't been fully implemented as of the time of the workshop, and thus, had not been fully tested.

According to Stringer, in the event of a catastrophe, the NRP calls for an accelerated provision of all federal assets during the first 48 hours following a disaster. Those assets, both human and other, will be directed to a federal mobilization site to avoid overwhelming the affected area until the quantity of federal resources needed for the response can be determined. Once federal and state assets arrive at the disaster site, they will assist and augment local assets. A Personnel Federal Official (PFO) will be charged with the task of ensuring that the coordination of those assets provides the full range of the nation's capabilities and that authority over the response effort remains with the local jurisdiction.

The NRP is designed to ensure that respondents from every level of government follow the basic incident command system and apply the basic principles of disaster medicine to triage and treatment of victims. Authorities will determine how to achieve the maximum good for the greatest number of victims, making it virtually impossible to maintain the traditional high-quality standards of care that currently exist in the day-to-day United States health care system.

National Disaster Medical System

In the event that an incident exceeds the capabilities of the local and state health care systems, the National Disaster Medical System (NDMS) serves as the lead federal agency for medical response under the National Response Plan, in collaboration with the United States Public Health Service's (USPHS) Commissioned Corps Readiness Force, the Department of Veterans Affairs (VA), and the Department of Defense (DoD). Operating within the U.S. Department of Homeland Security, Federal Emergency Management Agency, Response Division, Operations Branch, the NDMS coordinates medical response, patient evacuation, and hospitalization of victims of federally declared disasters, noted Stringer.

The entire NDMS system includes:

- Disaster Medical Assistance Teams (DMAT) are groups of professional and para-professional medical volunteers, supported by logistical and administrative staff, designed to provide medical care to disaster victims. DMATs are sponsored by a hospital, public health department, public safety agency, or local government. Sponsors recruit team members, arrange training, and coordinate team deployments. Teams deploy to disaster sites within 4 to 24 hours, with sufficient supplies to sustain their medical care responsibilities, in either fixed or temporary patient care sites, for a period of 72 hours.
- National Nurse Response Teams are trained to assist in mass chemo-prophylaxis, mass vaccination, and supplementation of the nation's nurse supply in the event of a weapon of mass destruction event.
- Disaster Mortuary Operational Response Teams (DMORT) are composed of private funeral directors, medical examiners, coroners, patholo-gists, forensic anthropologists, medical records technicians, finger-print specialists, forensic odontologists, dental assistants, x-ray techni-cians, mental health specialists, security and investigative personnel, and administrative support staff. DMORTs assist in establishing temporary morgues, victim identification, processing, preparation, and disposition of remains.
- Veterinary Medical Assistance Teams include clinical veterinarians, veterinary pathologists, animal health technicians, microbiologists/virologists, epidemiologists, and toxicologists, all of whom provide a range of surveillance activities and animal care treatments.
- National Pharmacy Response Teams assist in the distribution of prophy-laxis to Americans in the event of a bioterror attack or an emerging infectious disease epidemic that can be prevented with pharmaceuticals (DHHS, 2004b).
- National Medical Response Teams (NMRT) are three teams across the country that are equipped and trained to respond to a WMD event and provide victim decontamination and patient care to exposed victims. They carry their own personal protective equipment and a pharmaceutical stockpile to treat up to 5,000 victims. They have been mobilized in less than 4 hours two times since 2001.
- The Federal Coordinating Centers recruit hospitals to participate in the NDMS and, in the event that the system is activated, the FCCs coordinate the reception and distribution of patients being evacuated to areas not affected by the emergency.

According to Stringer, in the event of a mass casualty scenario, all 1,080 NDMS volunteers will be immediately activated, with the teams located closest

to the disaster mobilizing first, assuming that both air and ground transportation routes are available to transport the teams to the disaster site. The Department of Homeland Security's goal is to deploy 14 teams to the disaster site by the end of the first day. The entire system, less a few teams held back in the event of a secondary attack, could be deployed by the end of the third day.

As Stinger noted, the DMAT teams deployed to disasters would (1) establish alternate outpatient care facilities where victims can be treated with limited holding capacity (with the entire NDMS system deployed, team members can treat 224 inpatient and 4,500 outpatients per day in these facilities); (2) augment medical care in local outpatient facilities, treating 5,000 patients per day; (3) establish Casualty Collection Centers, collecting and assisting with the evacuation of patients to be treated in other parts of the country if the medical system near the disaster site is overwhelmed. With the entire NDMS system deployed, 4,200 patients can be evacuated to hospitals away from the disaster site; and (4) augment standard medical-surgery wards by sending DMAT teams to empty hospital wards to increase hospital surge capacity. Deployment of all DMAT teams would allow for treatment of 1,400 patients. While the activation of the NDMS would substantially increase the treatment capacity in the affected area, Stringer acknowledged that combined local, state, and federal resources would be severely overwhelmed in the event of a disaster involving 100,000 casualties.

Communication at the Department of Homeland Security

Since its inception, the Department of Homeland Security has been working to achieve widespread coordination by upgrading communications systems and equipment, as part of its new approach to protecting the country. In developing its new communication system, DHS employed the vertical and horizontal integration of assets that was previously described by William Raub, of the Department of Health and Human Services. New communications tools reach horizontally through all federal agencies and departments, as well as, vertically, to officials at the state, local, territorial, and tribal levels (DHS, 2004).

In addition to its color-coded Threat Condition, Information Bulletins, and Threat Advisories, the Department has created two new channels of communication—the National Infrastructure Coordination Center (NICC), created for the private sector, and the Homeland Security Information Network (HSIN), created for government agencies. The NICC allows industry representatives and individual companies to receive and provide information regarding specific threats and to be in constant communication with Department representatives during crises. The HSIN is a real-time collaboration system that provides emergency operations centers and governments, at every level, with the opportunity to share the same threat information so that all jurisdictions have the tools they need to make wiser decisions in securing their areas. Those two new communication systems support the Homeland Security Operations Center, a 24-hour, 7-days-a-week

communications center that aids the Department in monitoring activity throughout the nation. As Stringer observed, the Department's new communications systems are designed to stop a terrorist attack before it happens (DHS, 2004).

SUPPLIES

Utilizing cooperative agreement funding furnished to jurisdictions and hospitals, plans are being developed to strengthen the coordination and communication between hospitals and local, state, and federal agencies. In the event of a disaster, these detailed plans may call for drugs, vaccines, information, food, water, and other essential resources to be distributed among the public. Rapid community needs assessments must be completed to determine the amount of resources necessary, the members of the community in need, and the means to effectively distribute available resources to them, noted Stringer.

Rapid Needs Assessment

A rapid needs assessment is a low cost, statistically sound, population-based epidemiological tool that can be used following a disaster to provide emergency managers with accurate and reliable information about the needs of an affected community, as those needs change in the aftermath of a crisis. According to Carol Rubin, of the Centers for Disease Control and Prevention's National Center for Environmental Health, rapid needs assessments are adaptable to unique disaster situations and allow for evidence-based decisions and interventions.

Assessments are conducted as follows. First, a representative sample population is identified so that results can be extrapolated to the larger community; second, interview teams, composed of staff and volunteers from local, state, and regional health departments, administer community-specific surveys through face-to-face interactions with affected community members; finally, interviews, data entry, and data analysis are completed within 48 hours. According to Rubin, the "rapid" in rapid needs assessment refers to the speed and accuracy with which data are collected, processed, and utilized. Rubin further noted that rapid needs assessments have been successfully used in responding to hurricanes, floods, and ice storms. The information obtained through the assessment enables responders to comprehend the actual numbers of

> A rapid needs assessment is a low cost, statistically sound, population-based epidemiological tool that can be used following a disaster to provide emergency managers with accurate and reliable information about the needs of an affected community, as those needs change in the aftermath of a crisis.
>
> —*Carol Rubin*

resources needed, target specific warning messages to affected residents, and, in addition to identifying unmet health needs, assessments can provide real-time information about housing, mental health, and utilities services.

Following the initial assessment, it is important to periodically reassess residents' needs as relief activities progress. Needs may change over time, especially if families migrate into or out of the community. Periodic rapid needs assessments can also aid in the community's rebuilding process. When rebuilding infrastructure, Rubin suggested that interventions go beyond needs replacement, and, instead, aim for sustainable change.

To aid in analyzing the results of needs assessments, Samuel Wilson, of the National Institute of Environmental Health Sciences, suggested the development of a national database indicating Americans' baseline health status. Wilson noted that health officials' current understanding of the population's health status is insufficient and that the development of a baseline database will allow health officials to immediately understand the health impacts of a disaster following a rapid needs assessment.

Strategic National Stockpile

With results from the rapid needs assessment, responders can begin to distribute supplies to communities affected by the disaster. In the event of a national emergency, state, local, and private resources will be depleted rapidly; therefore, many supplies will come from the nation's Strategic National Stockpile (SNS).

In 1999, at the request of Congress, the Department of Health and Human Services and the Centers for Disease Control and Prevention began to invest significant financial resources in developing the capabilities to acquire, store, and distribute pharmaceuticals and medical supplies (e.g., intravenous fluids, airway maintenance supplies, and medical/surgical items). The Homeland Security Act of 2002 initially charged the Department of Homeland Security with managing the deployment of those assets, but, in March 2003, the stockpile became jointly managed by the Department of Homeland Security and the Department of Health and Human Services, under the Strategic National Stockpile title.

> The Strategic National Stockpile currently has a capacity of antibiotics to treat 13 million people for 60 days.
>
> —*William Raub*

SNS supplies can reach states and United States territories within 12 hours following a decision to deploy, thereby indicating that the stockpile is not to be used as a first response tool. Initial deliveries of assets would include 12-hour Push Packages, consisting of a broad spectrum of supplies that can supplement a region's existing stock until the specific needs of the community are determined.

If needed, additional shipments of products tailored to the nature of the disaster will follow within 24 to 36 hours.

The stockpile is located at 12 different sites, and, according to William Raub, it currently has a capacity of antibiotics to treat 13 million people for 60 days. Careful attention is paid to composition of the stockpile, based on biologic and/or chemical threats and the public's vulnerability. With many of the stockpile's assets consisting of antibiotics, vaccines, chemical antidotes, antitoxins, and life-support medications, the SNS Program must be extremely mindful of shelf-life and stock rotation.

States and territories can receive SNS assets through a governor's direct request to the CDC or the DHS. Once a decision has been made to deploy, assets will be loaded into trucks and/or commercial aircraft. It is then up to the state and local authorities, with assistance from the SNS Program's Technical Advisory Response Unit, to put the assets to use promptly (CDC, 2003).

CDC's ChemPak Program

As noted above, intelligence sources believe that terrorist groups may use nuclear, biological, chemical, or radiological weapons, potentially overwhelming the United States' response capabilities. Ideally, weapon of mass destruction events using unconventional agents can be prevented through the new, improved Homeland Security Operations Center; however, it is unlikely that all planned attacks can be thwarted. It is, therefore, the task of first responders to effectively prepare for an expedited mobilization of their resources to diminish morbidity, mortality, and destruction of structural infrastructure following a disaster.

While the Strategic National Stockpile is designed to provide states with pharmaceuticals and medical materiel within 12 hours, that would be an inadequate response time following an attack involving a nerve agent. Without prompt treatment, victims can suffer immediate nervous system failure and death. On a positive note, atropine sulfate, pralidoxime chloride, and diazepam are known antidotes to the harmful effects of chemical nerve agents. To distribute nerve agent antidotes in a timely manner, the Centers for Disease Control and Prevention has established the ChemPak program, a voluntary project that provides funds to cities and states to place nerve agent antidotes in monitored storage containers for immediate use in the event of a chemical emergency. Notwithstanding local storage, the SNS Program will maintain authority and control over the assets. ChemPak participating cities and states must agree to:

- Create sustainable plans for ChemPak project antidotes' dissemination, surveillance, and maintenance.
- Develop and implement strategies to maximize the shelf-life of the remedies, and abide by the provisions set forth by the Federal Drug Administration's Shelf Life Extension Program.

- Use the contents of the ChemPak containers only after it has been determined that an actual nerve agent release threatens public health.
- Develop a single state and/or city ChemPak program point of contact (POC).
- Determine the quantity of containers needed by first responders.
- Provide SNS program personnel with the address of each storage container for monitoring purposes and to ensure coordination of assets following the deployment of the SNS.
- Identify a licensed pharmaceutical or medical professional who will be responsible for accepting the delivery, storage, and safety of the ChemPak container contents (CDC, 2004b).

The assets stored in the 12 SNS sites and ChemPak program containers will help to ensure that adequate supplies can be deployed to disaster zones. According to William Raub, once these supplies reach the state or city drop-off site, emergency managers must determine an efficient method for distributing assets to each individual in need.

CDC's Cities Readiness Initiative

Prior to the 2001 anthrax attacks, Americans underestimated the likelihood of a national level bioterrorism attack, and, in so doing, overlooked some areas of the country where federal assets might be needed to assist in the response effort. To aid cities in successfully dispensing SNS assets following a bioterrorism attack or other large-scale disaster, the Department of Health and Human Services, in collaboration with the Department of Homeland Security, granted 27 million dollars of (FY) 2004 funds (Figure 3.1) to 21 selected cities as part of the Cities Readiness Initiative (CRI). The CRI is part of the federal government's considerable effort to increase the safety of Americans, demonstrated by over 130 million dollars of (FY) 2002 and (FY) 2003 funds distributed to state and local governments to strengthen their SNS distribution capabilities (CDC, 2004b).

Under the CRI, participating cities are to develop a template for administering supplies to affected residents, incorporating federal, state, and local government officials, as well as fire, police, emergency medical service, SNS, and United States Postal Service (USPS) personnel into the distribution effort. Traditionally, state facilities, other than hospitals, have been utilized to distribute chemoprophylaxis to residents who were potentially exposed to a

The Department of Health and Human Services has reached an agreement with the United States Postal Service to call upon their employees for direct residential delivery of antibiotics to those located in the disaster zone.

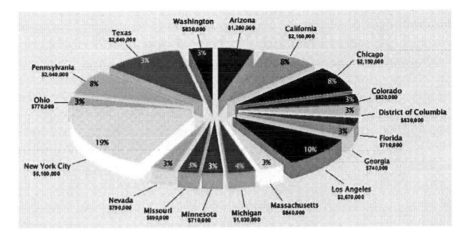

FIGURE 3.1 Distribution of CDC funding to cities and states for Cities Readiness Initiative activities. SOURCE: CDC, 2004.

chemical or biological agent. The CRI will enhance distribution by establishing a network of points of dispensing (PODs), staffed with well-trained volunteers and paid employees, who can provide information and recommendations to concerned residents, in addition to prophylactic antibiotics and antidotes. To further revolutionize dispersion methods, Raub noted that DHHS has reached an agreement with the USPS to call upon their employees on a voluntary basis for direct residential delivery of antibiotics to those located in the disaster zone. This cooperative effort will provide the speed of penetration into the community that will be necessary to control a public health catastrophe.

The results of this initiative will be to offer a consistent, nationwide approach for all jurisdictions to utilize to effectively distribute supplementary assets to the population. Once developed, verified, and exercised, local dispensing plans can help to save lives through timely delivery of SNS material during a naturally occurring or man-made public health emergency.

Future DHHS and DHS Preparedness Plans

In addition to the Cities Readiness Initiative, the Department of Health and Human Services and the Department of Homeland Security are partnering to enhance and upgrade field hospital supplies. According to Lew Stringer, since January 2004, multiple tractor trailers have been packed, each storing enough supplies for 150 beds. The trucks, stocked with items including: cots, blankets,

and portable toilets, are ready to be immediately mobilized, rather than waiting for preparation and packing. Funding has been provided for two field hospitals, and the planning process has begun. In the future, similar portable hospitals will be developed, further enhancing the United States' ability to respond to mass-casualty incidents. In addition, the federal government is purchasing transport vehicles for National Disaster Medical System volunteers and equipment to expedite the deployability of response teams. Once achieved, these new assets can increase the quality and speed of the response, thus reducing the magnitude and duration of the disaster's consequences.

STAFF

Following disasters, ample trained and credentialed volunteers are needed to assist in the medical response effort. According to Lew Stringer, even if all human resources from the USPHS Commissioned Corps Readiness Force, the Department of Veterans Affairs, and the NDMS are deployed simultaneously, the United States does not have an adequate contingent of medical professionals to stage an effective WMD response. The federal government's goal is to recruit and train 20,000 personnel, in addition to the existing VA, USPHS, Department of Defense, and NDMS staffs.

> The federal government's goal is to recruit and train 20,000 personnel, in addition to the existing VA, USPHS, DoD, and NDMS staffs.
>
> —*Lew Stringer*

Education and Training for Emergency Responders

The National Institute of Environmental Health Sciences (NIEHS) has been charged with the responsibility of training responders to protect themselves and their communities for the duration of the response effort. The Institute works to accomplish this task through its Worker Education and Training Program (WETP). Funded by the Superfund Amendments and Reauthorization Act of 1986, the WETP seeks to prevent work related harm by distributing grants to non-profit organizations to develop and deliver high quality occupational safety training and health education programs to workers exposed to hazardous materials and wastes.

Chemical waste sites can pose health and safety hazards to responders from unidentified chemical substances and the potential mixture of substances present. According to Samuel Wilson of the NIEHS, since 1987, approximately 80 awards have been granted to labor based groups, universities, and other academic institutions for the development of worker education and training models. Since that time, 1 million workers have benefited from the program's 14 million contact

hours of actual training designed to enhance the work practices and specialized technical skills of the workers who will be facing complex chemical responses.

Throughout the September 11 response in New York City, the Worker Education and Training Program monitored worker exposure, consulted on the development of a safety plan, and provided site safety training education and personal safety equipment to 4,000 clean-up workers at ground zero, noted Wilson. Workers were trained and certified in the use of their respirators to assure that they had some protection from hazardous fumes. Additionally, the program reestablished health and safety training programs for the FDNY, as many trained responders were, unfortunately, lost during the attacks.

Furthermore, according to Wilson, in the months following the terrorist attacks, the NIEHS funded many initiatives to evaluate New York City residents' health status, including: monitoring residents' personal exposure, collecting and analyzing air and dust samples, conducting respiratory health studies, initiating epidemiology studies, providing residents with exposure information and fact sheets, and advising clinicians about the related clinical conditions known to be associated with the disaster site.

While the training programs established for the September 11 response had an appreciable health effect on workers, the WETP is working to develop improved preparedness training for workers deployed in future responses. Wilson noted WETP's current efforts to:

- Establish training guidelines for emergency response and clean-up in the event of a WMD event.
- Provide a standardized framework for addressing public and worker monitoring, medical surveillance, protective equipment, and decontamination, according to the U.S. Occupational Safety and Health Administration's Hazardous Waste Operations and Emergency Response (HAZWOPER) guidelines.
- Identify safety equipment necessary for future responses in major urban centers.
- Continuously train workers in responding to new threats and emerging toxic materials, as scientific, medical, and technical aspects of disaster response tend to change rapidly.
- Create new horizontal and vertical partnerships between the public and private sectors at the national, state, and local levels.
- Develop peer-reviewed training materials, to ensure high-quality standards.

While the WETP's efforts will improve the safety of emergency responders as they complete their work, Wilson suggested the formation of a uniform national enterprise with the ability to partner with government and private sector training programs. Ideally, such a project would incorporate experts' emerging work on communications systems, training standards, and response protocols.

Management of Staff

While participants noted that organizations have begun to address training of staff and developing contingency plans for providing adequate staff during acute stages of crises, little work has focused on the management of staff. For example, a participant from the NYC Office of Emergency Management noted that Joint Commission on Accreditation of Healthcare Organizations (JAHCO) requires hospitals to train and perform exercises for a variety of scenarios, such as a plane crash, anthrax, and other similar situations. These exercises demonstrate that the health care providers are able to see an injury pattern or a particular disease, and they are able to access necessary information to initiate appropriate care. In a short-term crisis, this works well as staff will work through the situation. However, one participant questioned whether in a sustained event, such as those that could last for more than 24 hours, if the management and the support of the facility have considered the available human and supply resources. This means that staff would need to be given time off in order to be able to meet longer term staffing needs. Dr. Stringer echoed these concerns and said that his office has started engaging the local emergency management to look at how assets are managed when additional resources are not available. He further noted that his office is looking at some of the practices of the Veteran's Affairs hospitals and how these may be applicable to local hospitals, but he acknowledged that additional planning and study will have to be done.

SPACE

Along with improving response systems, acquiring adequate stockpiles of supplies, and recruiting, credentialing, and training response staff, it is just as important to ensure that sufficient physical space has been secured within which to successfully implement the medical response, observed Raub. Following a catastrophe, facilities will certainly be needed for patient care, mental health care, and treatment of minor injuries. In addition, should the affected area be deemed uninhabitable, separate venues may be needed for isolation, distribution of community prophylaxis, and evacuation of victims (Burstein, 2004).

Emergency managers have proposed transforming old hospitals, state facilities, and hotels into isolation sites, where temporary cots, blankets, and patient-care supplies could be assembled. As was noted above, in the event of an attack requiring mass chemoprophylaxis to prevent adverse health effects among the public, regional health officials and volunteers will form points of dispensing (POD) sites, noted Raub. POD sites must be located away from hospitals to prevent unnecessary overcrowding during a time when hospital facilities are likely to be incredibly overwhelmed. Some participants proposed using schools or other community meeting sites as potential points of dispensing. When choosing a site, emergency planners must consider those that are well-known to com-

munity members, as well as issues pertaining to security, adequate parking, and restroom facilities (Burstein, 2004).

The complex disasters that the United States may face in the future will require a carefully prepared, yet flexible, response. Preparedness and response efforts can be strengthened through the collective wisdom of generalists and specialists in the private sector, scientific, academic, and industrial communities, as well as government officials at every level—those who will ultimately coordinate, and be held accountable for, the events that occur before, during, and after disasters, concluded Wilson.

4

Practical Considerations of Emergency Preparedness[1]

PRACTICAL LOOK AT EMERGENCY PREPAREDNESS AND CRISIS MANAGEMENT: PROTECTING WORKERS AND CONTINUING ESSENTIAL SERVICES

Disasters can be a result of a natural agent, a terrorist act, or an industrial accident. Disasters can have impacts on businesses from both a personnel and an economic standpoint. Because many individuals are at work when disasters strike, it is even more imperative that businesses are a part of the planning for how to manage the impact of disasters and how to prevent them, said Jack Azar of Xerox, Inc. The interest in managing and preventing crises at Xerox started in December 1984 when a disastrous chemical release occurred in Bhopal, India, and 2,000 people were killed as a result of it.

Emergency preparedness at Xerox—which became especially acute due to the events of September 11, 2001—integrates several phases of response from a business perspective: emergency response, crisis management, and business continuity (Figure 4.1). The first phase, which is usually of a short duration, is the emergency itself. This may include a fire or an explosion at a plant or a facility. The initial management of the response to the emergency at a Xerox factory would involve the environmental health and safety committee, as well as the security department of the company. Their actions would be to protect the employees, property, surrounding communities, and the environment. The second phase, crisis management, is when the local event continues or increases in size as a result of uncertainty or crisis fall-out. Sometimes the emergency may last for weeks and cause concern in the public health sector. Thus, this phase needs

---

[1]This chapter was prepared by Dalia Gilbert from the transcript of the meeting. The discussions were edited and organized around major themes to provide a more readable summary and to eliminate duplication of topics.

to be handled by the senior management throughout the company. The CEO then would decide, based on recommendations from his team, how the company should proceed. The third phase is business continuity. If an accidental explosion occurs at a plant, for example, all operations at the plant are shut down. Sometimes it may be a critical operation to a company, and in some cases it may be the only particular site that has a product or material coming out of the plant to worldwide customers. It is important to know how, in case of a disaster, a business puts its employees back to work, resumes its operations, and keeps the customers happy and the economy thriving, said Azar. At Xerox continuity planning is in the hands of the operations group.

> It is important to know how, in case of a disaster, a business puts its employees back to work, how it resumes its operations and keeps the customers happy and the economy thriving.
>
> —*Jack Azar*

The processes of emergency management need to be formalized and standardized throughout the company. Xerox has their facilities and 60,000 employees worldwide, and even though the managerial level employees speak and understand English, it may be challenging to convey the standards to the entire workforce and to ensure that they are carried out. The approach Xerox used to address the challenges was to get together all of the major players from the worldwide facilities and to review the standards in a simple fashion so that the requirements are understood.

Putting the policy in practice, however, is not always easy, noted Azar. In 1999, Xerox started considering what it would do if they lost a site that produced a critical product and it was a sole site of production of that particular material.

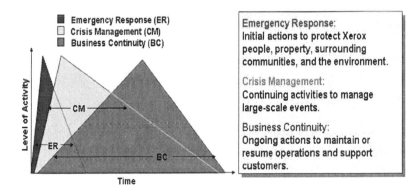

FIGURE 4.1 Establishing emergency management in industry. SOURCE: Xerox Corporation. Reprinted with permission.

The driving force was business continuity planning, and it was initiated across the company for consistent operation to ensure that the planning was done in India and Brazil the same way it was done in the United States. At the same time, the environmental health and safety department at Xerox updated their standard for local emergency preparedness.

Modifying Disaster Planning as a Result of the 2001 Terrorist Acts

On September 11, 2001, Xerox had approximately 100 employees in the World Trade Center who were customer service representatives; the company also had business operations located in both WTC buildings. Xerox lost 2 employees and about 100 survived, but it took several days before the company had an accounting of its employees, noted Azar. The experience taught the company a valuable lesson, and it was still in the process of solving the issue of getting its customers back in operation when the anthrax threat began. The threat affected Xerox because the company has 400 operations across the country, primarily in large cities, that do mail sorting for many companies. Some of the facilities were in New York, New Jersey, Washington, D.C., and Florida, where the hot spots for the anthrax operation occurred. Due to the media reports and the constant handling of mail, the employees at Xerox were concerned about their health. In an effort to protect them, the company requested guidance from upper management and assembled a mail safety team. This team followed and tracked the information released by the CDC and the U.S. Postal Services. After the CDC advisory, Xerox made it mandatory to equip its employees with disposable respirators and gloves. The respirators selected to protect from anthrax spores are N95 type. However, it was very difficult to obtain them because they were in high demand. The Postal Services alone bought about 4 million respirators in the course of two weeks, and it took the procurement and environmental health and safety departments at Xerox about two weeks to locate available supplies for 1,000 people.

Shelter in Place

The other challenge that Xerox had was to include shelter in place planning while creating an emergency management plan. Xerox has about 7,500 employees at its Webster facility in New York State. They work within 7 miles of a nuclear power plant. Since Xerox is the largest commercial employer within the 10 mile radius from the plant, it was asked to develop a shelter in place plan in case of terrorism or an accidental release from the facility. At the same time, a crisis management team at the senior levels in the company was created. That team reports directly to the chief of staff, who is in constant communication with the CEO. The team includes operations, health and safety, and security employees as well as public relations, employee communications, and human resources

employees. It took Xerox a year to develop the proper employee communication for a shelter in place plan because previous evacuation alarm systems, used in cases of emergency, forced people to go outside. Thus, Xerox had a challenge to work out a new mode of communicating with its employees. Today, Xerox has two drills a year; one is a fire drill for evacuation where the tone of alarm is very loud, the other is a shelter in place drill where there is a different tone of alarm followed by a PA system communication. Shelter in place procedures were used at the Webster facility in December 2003 when the company had an on-site shooting related to an armed robbery at the Federal Credit Union.

Need for Additional Coordination

Coordination and flow of information is a critical need for industry. There is little coordination within industries, with the majority of the information sharing occurring through interactions with various governmental agencies. Companies do communicate in order to benchmark operations for renewing business, but the effort is not systematic and does not focus on emergency management. Similarly, additional communication needs to occur between industry and government agencies, noted Azar. Health and safety teams have difficulty obtaining accurate information. Sometimes information on government web sites is contradictory, and it is difficult to talk to someone to obtain accurate information. In the case of anthrax, unless one knew someone at the CDC, it was difficult to obtain good advice. Azar concluded by referring to the need for a more open process as there are more than 100 million people who work in the sector and very often an emergency happens during work hours.

NGO'S ROLE IN PUBLIC CAPACITY BUILDING: THE AMERICAN RED CROSS

The American Red Cross is an organization that is directly engaged in the neighborhoods where people live. Unlike many federal agencies, the American Red Cross is not a science agency; it does not have medical experts, seismologists, meteorologists, or hydrologists to conduct research, said Rocky Lopes of the American Red Cross. However, it does have many people who provide a great variety of accurate, appropriate, and sensitive information to the public. The American Red Cross collaborates extensively with a number of agencies in order to provide accurate and understandable information, said Lopes.

The American Red Cross works very closely at the national level to inform the public of appropriate actions. Lopes noted that some of the existing emergency preparedness information that can be found throughout the country is not based on science; it is folklore that interferes with people's understanding about what to do. For example, some people think that in case of a hurricane one should cover only the windows in the front of one's home, but hurricane winds

come from all directions, not just the front of a building. Thus, the American Red Cross works closely with FEMA, the Department of Homeland Security, and the National Weather Service to convey the same message so that wherever people turn in their process of verification, they get consistent advice.

Emergency planners need to enable people to understand both what can happen and what actions they can take, and to understand that people are looking for information from a variety of sources. While federal government agencies have substantial information on emergency preparedness, emergency planners need to take into consideration that there are many people in the United States who do not turn to government for information, or trust government, asserted Lopes.

Generally, people trust organizations that provide credible, reliable, believable, and meaningful information to them, asserted Lopes. This means that some NGOs are well-positioned with certain segments of the public and therefore have a greater reach and level of penetration within that segment. But some agencies and organizations need to get over the perception of the ownership of message, noted Lopes. When it comes to emergency preparedness, it should not be the Red Cross message, a government message, or a church

> People shop around for information and compare one organization's message with another.
>
> —*Rocky Lopes*

message. It should be the same message coming from all the organizations, said Lopes. It is by far more important that people get the message rather than the identity of the deliverer of the message. This can sometimes be challenging within the political arena, especially in Washington, D.C., said Lopes.

Further, he noted that repetition of messages reinforces and inspires action. It is not enough to tell the public once that they need to be prepared and to expect them to be prepared. People engage in verification. If the same message is provided by multiple organizations it becomes more credible to the public (Mileti, 1999). This data suggests a need for the American Red Cross to collaborate with other organizations and a need to ensure consistent messaging. Even though the American Red Cross is not a scientific organization, it relies on science from other organizations and translates that knowledge into meaningful information for the public. The Red Cross collaborates with the National Disaster Education Coalition (NDEC), which is composed of 21 federal agencies and national nonprofit organizations. Prior to September 11, 2001, NDEC consisted of only eight organizations: the Red Cross, FEMA, Weather Service, USGS, National Fire Protection Association, International Association of Emergency Managers, Institute of Business and Home Safety, and the U.S. Department of Agriculture Extension Service. Subsequent to the September 2001 events, more organizations became involved because of the need to make the messages more consis-

tent. The organizations meet monthly, catalog their information, validate it through research and publish it on the web site www.disastereducation.org, as well as through the web sites of the Red Cross, FEMA, NOAA, and others. The information available on the web sites is designed for those who communicate with the public: educators, web site designers, brochure writers, newsletter article writers, and others. The information can be tailored depending on the target group because the more local and relevant it is made, the more likely it is to get a response from the public and to build public capacity, noted Lopes.

> If the same message is provided by multiple organizations it becomes more credible.
>
> —*Rocky Lopes*

Thus, the most critical thing in building the capacity among the public is to provide information consistently. When the information is put out in a variety of venues, many people will put it to use, thereby reducing the potential for death, injury, and property damage in all types of future disasters, concluded Lopes.

DISPLACED CHILDREN AND THE COMMUNITY

When emergency planners review different emergency scenarios, they usually base their scenarios on adult, educated, healthy people. These plans, however, may not be useful for socially vulnerable groups such as the handicapped, immobile elderly, immigrants with limited English skills, or children during disasters, said J. R. Thomas of the Emergency Management Office in Franklin County, Ohio. Thomas used issues that related to children during a disaster of large proportions to begin a discussion of the complexity of ensuring that the needs of these socially vulnerable groups are met.

> Children cannot be treated as small adults because their age, cognitive skills, and comprehension of the surroundings is different.
>
> —*J. R. Thomas*

Children are a particularly vulnerable population. They may be able to walk and talk, but they cannot be treated as small adults because their age, cognitive skills, and comprehension of the surroundings are different. There are many challenges that emergency planners must consider in situations involving children. To ensure that children's medical, legal, physical, and psychological needs are met, these issues need to be discussed and planned for in advance, said Thomas.

Medical Issues

Under ideal circumstances, if a child has a medical emergency, he or she goes to a local children's hospital where a pediatrician examines them. However, in the case of a disaster there may be more than 1,000 children admitted to a hospital in a day and not all of them will be able to go to a children's hospital. Some will need to go to a regular emergency department to be treated. This may be problematic because the regular emergency department may not have adequate equipment for children, noted Thomas. For example, a typical respirator as well as some surgical tools will not work for a child. Emergency medical services and urgent care centers will need to have better access to child size equipment.

Physical Issues

Further, if a child is lost, a police officer takes him or her to Children's Services and Children's Services find them temporary housing. During a disaster, because of the large number of displaced children, it may take several days before a placement is found for all the children in need of housing. During those several days the children will need personal hygiene equipment and nutrients; small children may need diapers and formula. Decisions may need to be made as to whether it is better to have 50 children in a gymnasium or to have 20 children in a room in somebody's house, noted Thomas. Temporary placement in a foster home might be a better solution than putting children in a shelter, but then social services may be unable to find foster homes for large numbers of children. An emergency management office needs to be cognizant of the facts that other things have to take place in a situation where people, especially children, need to be moved and it is very important to think of this process now and not wait until it becomes an actual situation, said Thomas.

Legal Issues

Housing for a significant number of displaced children may result in legal implications, especially when normal processes and access to information may be disrupted. Parents need to know where to look for their child in case of an emergency displacement, while local officials will need to define the parameters for transferring legal custody. If 1,000 children have to be placed somewhere, emergency planners and people responsible for the children's safety need to know if the person who comes to pick them up is really a next of kin, or somebody who has legal authority, for example, a custodial parent. Another legal issue is whether it is permissible to release a child to a non-custodial parent without a court order. If a child is picked up by someone other than a parent, that person would need an approval and their background would need to be checked as well. However, it is almost impossible to have a court hearing for each indi-

vidual case if there are 50 or more children in question, and it is hard to decide whether a probate court, a juvenile court, or a magistrate's court should process the cases. Furthermore, if children have to be moved out of a downtown area, and a courtroom is closed, emergency planners need to think of where a hearing would be held and whether it would be possible to set up courtrooms in convention centers or other large venues, noted Thomas. The judicial systems will need to have contingency plans in place to provide expeditious handling of cases and to determine when flexibility of legal standards should be explored.

Psychological Issues

Emergency planners need to think about mental health capabilities as well as long-term care in case of post-traumatic stress disorder in children. It is very traumatic for a child to lose a parent or both parents in an incident, and it is essential that emergency management departments find a way to coordinate social workers, pediatric psychologists and, if needed, psychiatrists to help children in distress, noted Thomas.

Thomas concluded by stating that the ultimate goal of every emergency management organization is to reunite children with their parents or relatives as quickly as possible. Therefore, communication between organizations that handle emergency situations is very important. Organizations such as NGOs and government agencies need to work together and plan ahead to identify the areas where children in distress are going to be taken; they need to ensure that transportation is available and that children are accounted for, concluded Thomas.

WRAP-UP

The discussions of the workshop were quite sobering on the health issues and other challenges that the United States and other countries face during a time of disaster. As destructive as natural disasters such as tornados are, they can be addressed because their intrinsic hazards do not change from disaster to disaster. Terrorist events, however, are difficult to prepare for and defend against because terrorists can change their method of operations. Therefore, an integration of disciplines, especially for public health and emergency responders, needs to be in place in order to meet the challenges and to be effective during a time of crisis, noted Bernard Goldstein, Graduate School of Public Health at the University of Pittsburgh.

As the United States continues to plan for responding to disasters, research and training must guide the effort. Many people believe that once a terrorist event is concluded, the threat from terrorism is reduced. This is a misconception because there are likely to be more terrorist attacks in the future. In order to keep up with potential threats, emergency planners have to focus on training. The

field will also have to be able to systematically evaluate their response with better tools, asserted Goldstein.

Communication was a central theme during the workshop, and ranged from communication capacity at the local level to the need for more research communication. During the workshop, communication at the national level was emphasized; however, Goldstein noted that the majority of people in the country obtain most of their information from local sources. People do not turn to CNN or the CDC, but rather to the local health commissioner and the local TV and radio stations for information.

Local health departments are traditionally very small in the United States. Often, during a time of crisis, a local health department is busy attending to the health needs of affected people and does not have the time to develop an effective communication strategy. Goldstein suggested that there is a great need for a communication surge capacity and to have knowledgeable people to answer the phone, as well as to ensure that messages are consistent for the media and the public.

Additionally, there is a need for more work on the science of communication, observed Goldstein. There is a pervasive belief that if one has the right information, then everyone will understand the risk and take the right action. CDC is therefore emphasizing the necessity for more research in risk communication that would provide a better understanding of how various groups process messages from the scientific community.

The second theme during the workshop was the call for building capacity, which will have to occur through partnerships between NGOs and the government, public and private sector organizations, and federal and local entities. Goldstein emphasized the critical partnership between federal and local governments because in the United States people rely heavily on local government. In contrast to some European countries such as France, the United States takes a decentralized approach to emergency management, with local management in charge during times of disaster. This is not likely to change, so it is important to find ways to strengthen the local/federal partnership and increase intergovernmental cooperation.

Goldstein concluded that, despite all the challenges we face, it is obvious that we have come a long way toward preparing for disasters since September 11. Yet we have so much further to go. Additional progress will not be easy; but it is reassuring that we know so much more today than we did before.

References

Baker S, Runyan C. 2002 (May 12). *William Haddon, Jr., His Legacy.* Lecture presented at the 6th World Conference on Injury, Montreal, Canada. [Online]. Available: http://www.sph.unc.edu/iprc/document/p1_files/frame.htm [accessed July 1, 2004].

Burstein JL. 2004 (February 10). *Bioterrorism 2004.* Lecture presented at the Harvard School of Public Health, Boston, MA.

Careless J. 2004. Blackouts won't stall responses in Windy City. [Online]. Available: http://www.emsmagazine.com/ffr/ffrmar04001.html [accessed July 28, 2004].

Covello VT, Sandman PM. 2001. Risk communication: Evolution and revolution. In: Wolbarst A, ed. *Solutions to an Environment in Peril.* Baltimore, MD: John Hopkins University Press.

Kellermann AL, Todd KH. 1996. Killing heat. *New England Journal of Medicine* 335:126–127.

Mileti D. 1999. Disasters by Design: A Reassessment of Natural Hazards in the United States. Washington, DC: Joseph Henry Press.

National Research Council (NRC). 2002. Making the Nation Safer: The Role of Science and Technology in Countering Terrorism. Washington, D.C.: National Academy Press.

Ott WR. 1990. Total Human Exposure: Basic Concepts, EPA Field Studies, and Future Research. *Journal of the Air and Waste Management Association* 40:966–975.

Payment, P., Siemiatycke, J., Richardson, L., Renaud, G., Franco, E., and Prevost, M. 1997. A Prospective Epidemiologic Study of the Gastrointestinal Health Effects Due to the Consumption of Water. *International Journal of Environmental Health Research* 5–32.

Pollard WE. 2003. Public Perceptions of Information Sources Concerning Bioterrorism Before and After Anthrax Attacks: An Analysis of National Survey Data. *Journal of Health Communication.* Supplement 1:93–103.

Rasool I, Baldi M, Wolter K, Chase TN, Otterman J, Pielke Sr. RA. 2004. August 2003 heat wave in western Europe: An analysis and perspective. Available [On-line]. http://blue.atmos.colostate.edu/publications/pdf/R-279.pdf [accessed July 28, 2004].

Staniland N. 2001. Injury prevention and control: Understanding the issues and making a difference. *International Journal of Trauma Nursing* 7:67–69.

Tubbesing SK, ed. 1994. The Loma Prieta, CA, Earthquake of October 17, 1989-Loss Estimation and Procedures. USGS Professional Paper 1553-A. US Government Printing Office.

Tierney KJ, Lindell MK, and Perry RW. 2001. Facing the Unexpected: Disaster Preparedness and Response in the United States. Washington, DC: Joseph Henry Press.

United Nations Educational, Scientific and Cultural Organization (UNESCO). 2003. Water for People, Water for Life. Paris, France: UNESCO Publishing.

U.S. Centers for Disease Control and Prevention (CDC). 2002. *About Extreme Heat.* [Online]. Available: http://www.bt.cdc.gov/disasters/extremeheat/about.asp [accessed July 23, 2004].

U.S. Centers for Disease Control and Prevention (CDC). 2003. *Strategic National Stockpile.* [Online]. Available: http://www.bt.cdc.gov/stockpile/index.asp [accessed July 6, 2004].

U.S. Centers for Disease Control and Prevention (CDC). 2004a. *Communication at CDC: emergency and risk communication.* [Online]. Available: http://www.cdc.gov/communication/emergency/erc_overview.htm [accessed July 1, 2004].

U.S. Centers for Disease Control and Prevention (CDC). 2004b. *Continuation Guidance for* Cooperative Agreement on Public Health Preparedness and Response for Bioterrorism—*Budget Year Five.* [Online]. Available: http://www.bt.cdc.gov/planning/continuationguidance/index.asp [accessed July 8, 2004].

U.S. Department of Health and Human Services (DHHS). 2004a. *Guidelines for Bioterrorism Funding Announced.* [Online]. Available: http://www.hhs.gov/news/press/2003pres/20030509.html [accessed July 6, 2004].

U.S. Department of Health and Human Services (DHHS). 2004b. *National Disaster Medical System.* [Online]. Available: http://ndms.dhhs.gov [accessed July 8, 2004].

U.S. Department of Homeland Security (DHS). 2003. *Initial National Response Plan.* [Online]. Available: http://www.uscg.mil/hq/g-o/g-opr/NRP%20Initial%20signed%2022Oct03.pdf [accessed July 6, 2004].

U.S. Department of Homeland Security (DHS). 2004. *Statement by Secretary Tom Ridge before the National Commission on Terrorist Attacks Upon the United States.* [Online]. Available: http://www.rwb.gov.edgesuite.net/dhspublic/interapp/testimony/testimony_0026.xml [accessed July 8, 2004].

U.S. Environmental Protection Agency. 2002. The Clean Water and Drinking Water Infrastructure Gap Analysis. Washington, D.C.

U.S. Executive Office. 2003. *Homeland Security Presidential Directive 5/HSPD-5.* [Online]. Available: http://www.whitehouse.gov/news/releases/2003/02/20030228-9.html [accessed July 6, 2004].

U.S. Geological Survey. 2004. Estimated Use of Water in the United States in 2000. http://water.usgs.gov/pubs/circ/2004/circ126B/htdocs/text-total.html

Water Infrastructure Network. 2000. Clean Safe Water for the 21st Century. Washington, D.C.

Wright R. 2005. Environmental Science. Englewood Cliffs, NJ: Prentice-Hall.

Zimmerman R. 2004. Water Chapter 5 in R. Zimmerman and T. Horan, eds., *Digital Infra-structures, Enabling Civil and Environmental Systems through Information Technology,* London, UK: Rutledge.

Appendix A

Workshop Agenda

**PUBLIC HEALTH RISKS OF DISASTERS:
BUILDING CAPACITY TO RESPOND**

Co-Sponsored by
The Disasters Roundtable
and
The Roundtable on Environmental Health Sciences, Research and Medicine
The National Academies, Room 100
500 Fifth Street, NW, Washington, DC 20001

JUNE 22, 2004

8:30 a.m.   Welcome, Introductions, and Workshop Objectives
William Hooke, Chair, Disasters Roundtable
Paul Rogers, Chair, Roundtable on Environmental Health Sciences,
Research and Medicine

SESSION 1: LINKING HAZARDS AND PUBLIC HEALTH (PART 1)

Moderator: Yank Coble, President-Elect, World Medical Association

8:50 a.m.   Communicating Science to the Public
Julie Gerberding, Director, Centers for Disease Control and Prevention

9:10 a.m.   Health Effects Following Terrorism
Lynn Goldman, Professor, Johns Hopkins Bloomberg School of Public
Health

9:30 a.m.   Questions and discussion

10:00 a.m. Break

SESSION 1: LINKING HAZARDS AND PUBLIC HEALTH (PART 2)

Moderator: Joseph Barbera, Co-Director, Institute for Crisis, Disaster, and Risk Management, George Washington University

10:20 a.m.   Disaster–Public Health Nexus[1]
             Linda Bourque, Associate Director, Center for Public Health and
             Disasters, University of California, Los Angeles

10:40 a.m.   Social and Health Effects During Heat Waves
             Eric Klinenberg, Assistant Professor, New York University

11:00 a.m.   Infrastructure Loss as a Public Health Risk
             Rae Zimmerman, Director, Institute for Civil Infrastructure Systems,
             New York University

11:20 a.m.   Complex Disasters and Public Health
             Jean-Luc Poncelet, Chief, Emergency Preparedness, Pan American
             Health Organization

11:40 a.m.   Questions and discussion

12:10 p.m.   Lunch break

PREPARING FOR THE FUTURE: CAPACITY BUILDING AND LESSONS LEARNED (PART 1)

Moderator: Ann-Margaret Esnard, Cornell University

1:55 p.m.   Which Part of "Emergency" Didn't You Understand?
             William Raub, Principal Deputy Assistant Secretary for Public Health
             Emergency Preparedness, Department of Health and Human
             Services

2:15 p.m.   Public Health Monitoring and Training Needs
             Samuel Wilson, Deputy Director, National Institute of Environmental
             Health Sciences

---

[1] Linda Bourque was unable to be present at the workshop; however, her presentation is available on the Disasters Roundtable website.

2:35 p.m. Capacity Building to Respond
Lew Stringer, Senior Medical Advisor, Department of Homeland Security

2:55 p.m. Rapid Assessment of Health Effects During Disasters
Carol Rubin, Chief of the Health Studies Branch, National Center for Environmental Health, Centers for Disease Control

3:15 p.m. Questions and discussion

3:45 p.m. Break

PREPARING FOR THE FUTURE: CAPACITY BUILDING AND LESSONS LEARNED (PART 2)

Moderator: Ellis M. Stanley, Sr., Manager, City of Los Angeles, Emergency Preparedness Department

4:05 p.m. Practical Look at Emergency Preparedness and Crisis Management: Protecting Workers and Continuing Essential Services
Jack Azar, Senior Vice-President, Health and Safety, Xerox Corporation

4:25 p.m. NGO's Role in Capacity Building of the Public
Rocky Lopes, Manager, Community Disaster Education, American Red Cross

4:45 p.m. Displaced Children and the Community
J. R. Thomas, Director, Emergency Management Office for Franklin County, Ohio

5:05 p.m. Questions and discussion

5:35 p.m. Wrap-Up
Bernard D. Goldstein, Dean School of Public Health, University of Pittsburgh

6:00 p.m. Adjourn

Appendix B

Speakers and Panelists

Jack Azar
Vice President, Environment, Health
  and Safety
Xerox Corporation

Joe Barbera
Co-Director, Institute for Crisis,
  Disaster and Risk Management
George Washington University

Linda Bourque
Associate Director, Center for Public
  Health and Disasters
University of California, Los Angeles

Yank Coble
President-Elect
World Medical Association

Ann-Margaret Esnard
Cornell University

Julie Gerberding
Director
Centers for Disease Control and
  Prevention

Lynn Goldman
Professor, Department of
  Environmental Health Sciences
Bloomberg School of Public Health,
  Johns Hopkins University

Bernard Goldstein
Dean, School of Public Health
University of Pittsburgh

William Hooke
Chair
Disasters Roundtable

Eric Klinenberg
Assistant Professor, Department of
  Sociology
New York University

Rocky Lopes
Manager, Community Disaster
  Education
American Red Cross

Jean Poncelet
Chief, Emergency Preparedness
Pan American Health Organization

William Raub
Acting Assistant Secretary
Health and Human Services

Paul Rogers
Chair
Roundtable on Environmental Health
    Sciences, Research and Medicine

Carol Rubin
Chief, Health Studies Branch
National Center for Environmental
    Health
Centers for Disease Control and
    Prevention

Ellis Stanley
Manager
City of Los Angeles, Emergency
    Management Department

Lew Stringer
Senior Medical Advisor
Department of Homeland Security

J.R. Thomas
Director
Emergency Management Agency for
    Franklin County

Samuel Wilson
Deputy Director
National Institute of Environmental
    Health Sciences

Rae Zimmerman
Director, Institute for Civil Infrastructure Systems
Robert F. Wagner Graduate School of
Public Service

Appendix C

Workshop Participants

Dori Ackerman
GRS Solutions.com

Holly Adams
Eastern Correctional Institute

Stephen Ambrose
NASA

Thomas L Anderson
Multidisciplinary Center for
    Earthquake Engineering
    Research, SUNY at Buffalo,

Linda Arapian
Children's National Medical Center

Stacey Arnesen
National Library of Medicine

Joan Aron
Science Communication Studies

John Babb
Office of the Surgeon General

Alina Baciu
Institute of Medicine

Anne Bailewitz
Baltimore City Health Department

Barbara Bailey

Isha Bangura

Lauren Barsky
Disaster Research Center, University
    of Delaware

Samuel Benson
NYC Office of Emergency
    Management

Frank Best

Benita Boyer
Loudoun Health District/VDH

Nicole Brown
Maryland Department of Health and
    Mental Hygiene

Michael Bryce
Department of Health and Human
    Services

Roger Bulger
Association of Academic Health
　Centers

Duane Caneva
U.S. Navy

Michael Castrilla
Office for Domestic Preparedness

Francine Childs
Baltimore City Health Department

Angela Choy
U.S. General Accounting Office

John Clizbe
City of Alexandria

Kristin Cormier Robinson
National Emergency Management
　Association

William Cumming
Vacation Lane Group

Tom Davy
Bureau of Medicine and Surgery

Daniel Dodgen

Julie Egermayer
Office of Space Science and
　Application

Sharon Eiler
Carroll County Health Department

Robert Ek
University of Delaware

Debra Evans
Medical Society of the District of
　Columbia

Lynne Fairobent
American Association of Physicists in
　Medicine

David Feary
National Research Council

Lauren Fernandez
Office for Domestic Preparedness

Timothy Foresman
International Center for Remote
　Sensing Education

Margaret Fowke
National Weather Service

Erin Fowler
Department of Health and Human
　Services

Leslie Friedlander
Immigration and Customs
　Enforcement (ICE)

Kenneth Friedman
U.S. DOE Office of Energy Assurance

MaryAnn Gahhos
Peace Corps

Harry W. Gedney
National Park Service

Don Geis
Geis Design-Research Associates

Jerry Gillespie
University of California, Davis

Gabriela Gonzalez
DC Government, Department of
Health

Kay Goss
Electronic Data Systems Corporation

Sandra Gregory
Maryland Department of Health and
Mental Hygiene

Rachel Gross
MDB Inc.

Mary Gunnels
National Highway Traffic Safety
Administration, USDOT

Dan Hanfling
Inova Health System

John Hicks
Federal Drug Administration

John Hoyt
Department of Homeland Security

Kathi Huddleston
George Mason University

Chip Hughes
National Institute of Environmental
Health Sciences

Barbara Jasny
Science/AAAS

Jeanette Jenkins
Community Health Administration

Peter Jensen
Children's Hospital

Carole Kauffman
Anne Arundel County Department of
Health

Rachel Kaul
Maryland Department of Health and
Mental Hygiene

Mark Keim
Centers for Disease Control and
Prevention

Edward Kennedy
U.S. NORTHCOM

Kristi Koenig
Office of Public Health and Environmental Hazards

Elissa Laitin
Arlington County Public Health

Elizabeth Lemersal
Federal Emergency Management
Agency

Edward Lennard
BlueCross BlueShield Association

Shulamit Lewin
International Center to Heal Our
Children

Cynthiana Lightfoot
The Washington DC EMS Association

Sarah Lister
Congressional Research Service

Cindy Lovern
Emergency Preparedness and
    Response

Gary Lupton
Fairfax/Falls Church Community
    Services Board

Anthony Macintyre
George Washington University

PJ Maddox
George Mason University

James Madsen
U.S. Army Research Institute of Medi-
    cal Research (USAMRICD)

Lisa May
Emergency Preparedness & Response

Ruth McDonald

Tom McGinn
Department of Health and Human Ser-
    vices

Carolyn McMahon
Atmospheric Policy Program

Kayvon Modjappad
University of Alabama at
    Birmingham Schools of Medicine
    and Public Health

James Moore
U.S. Public Health Service

Maurice Morales
U.S. Naval Reserve

Van Morfit
Health Resources and Services
    Administration

Jonas Morris
Health Services Development Inc.

Vladimir Murashov
National Institute of Occupational
    Safety and Health

Ahmad Naim
Thomas Jefferson University

Joanne Nigg
University of Delaware

Jennifer Nuzzo
Center for Biosecurity, University of
    Pittsburgh Medical Center

Daniel O'Brien
Office of the Attorney General

Sean O'Donnell
Metropolitan Washington Council of
    Governments

Michele Orza
U.S. G.A.O.

Cindy Parker
Johns Hopkins Center for Public
    Health Preparedness

Alan Perrin
U.S. Environmental Protection
    Agency

Alan Roberson
American Water Works Association

Havidán Rodríguez
University of Delaware

Carol Rubin
Centers for Disease Control and
    Prevention

Elizabeth Ruff
Carroll County Health Department

Carla Russell
Disaster Research Center, University
    of Delaware

Debbie Saylor
Carroll County Health Department

Rhonda Scarborough
U.S. Government Printing Office

Randolph Schmid
The Associated Press

John Scott
Center for Public Service
    Communications

Kurt Seetoo
Prince George's County Health
    Department

E. Marie Simpson
Anne Arundel County Department of
    Health

Christa-Marie Singleton
Baltimore City Health Department

Adrienne Smith
Child and Adult Neurology

Danielle Smith
National Research Council

David Speidel
Queens College, CUNY

Cathy St. Hilaire
Sciences International

Eugene Stallings
National Hydrologic Warning Council

J. Starlin
University of North Carolina Health
    Care System

Joe Steller
National Institute of Building Sciences

Jessica Strong
Prince George's County Health
    Department

Patricia Swartz
Maryland Department of Health and
    Mental Hygiene

Kathy Sykes
U.S. Environmental Protection
    Agency

Richard Sylves
University of Delaware

Nate Szejniuk
University of North Carolina Health
    Care System

Astrid Szeto
Food and DrugDrug Administration

Judith Theodori
Johns Hopkins Applied Physics
    Laboratory